Praise for *Sex on Earth*:

A writer who blends professional
charm, wit, and a cockeyed sense o
through nature's red-light district c

*Sex on Earth*a is a refreshingly self-aware exploration of the most intimate moments in nature and how the incredible variety of life has led us to frame our own thoughts about this near-ubiquitous biological drive. *National Geographic*

… does an admirable job of separating the scientific from the smutty … highly entertaining and informative. *BBC Focus*

Terrifically entertaining … Howard writes with self-deprecating charm and genuine enthusiasm … hubba hubba. *The Boston Globe*

Quirky, irreverent and endlessly fascinating.
Melissa Harrison, author of *Clay*

The infectious optimism of his fascination with an Earth full of beings doing exactly what they need to go on gives a comforting sense that everything is right with the world. *Publisher's Weekly*

Howard has written with Bill Bryson–like wit about the sex of pandas, dinosaurs, frogs, flamingos, hedgehogs, insects, and other creatures. Readers may laugh while they learn. *Booklist*

An enjoyable romp through the stories behind some of the world's most carnal critters. Ben Garrod, TV presenter

A wonderful picture of the driving force behind all animal life on earth: sex … this entertaining and very informative read will make you raise your eyebrows out of sheer innocent wonder-ment. *The New Indian Express*

A NOTE ON THE AUTHOR

Jules Howard is a zoologist, nature writer, blogger and broadcaster. He writes on a host of topics relating to animal life and appears regularly in *BBC Wildlife Magazine* and on radio and TV, including the BBC's *The One Show*, as well as *BBC Breakfast* and Radio 4's *Today* programme.

Like many zoologists, Jules has worked with a number of wildlife organisations, including BirdLife International, the RSPB and the Zoological Society of London. Unlike many zoologists, Jules spent three years of his life running a frog phone-line, answering questions from concerned members of the public (mostly about frogs).

SEX ON EARTH

A celebration of animal reproduction

Jules Howard

Extract on pp. 194–195 from *A Monograph of the Slug Mite – Riccardoella limacum
(Schrank.)* by F. A. Turk and S-M. Phillips (*Proceedings of the Zoological Society
of London*, 1976, vol. 115, 448–472) reprinted by permission of John Wiley & Sons
(onlinelibrary.wiley.com). Extract on p. 196 from *Studies on the life history
of Riccardoella limacum (Schrank)* (Acari-Trombidiformes) by R. A. Baker
(*Journal of Natural History*, 1970, vol. 4, 511–519), reprinted by permission
of the publisher (Taylor & Francis Ltd, *http://www.tandf.co.uk/journals*).
Quote on p. 237 from *King Solomon's Ring* by Konrad Lorenz (1952),
reprinted by permission of Taylor & Francis Ltd.

Bloomsbury Sigma is an imprint of Bloomsbury Publishing Plc
50 Bedford Square
London
WC1B 3DP

www.bloomsbury.com

Bloomsbury is a trademark of Bloomsbury Publishing Plc

Bloomsbury Publishing: London, New Delhi, New York and Sydney

A CIP catalogue record for this book is available from the British Library

ISBN (paperback) 978-1-4081-9343-3
ISBN (ebook) 978-1-4081-9342-6

10 9 8 7 6 5 4 3 2 1

Illustrations by Sam Goodlet
Typeset by Mark Heslington Ltd, Scarborough
Printed and bound by CPI Group (UK) Ltd, Croydon, CR0 4YY

Bloomsbury Sigma, Book One

For Emma
(and Tian Tian and Yang Guang)

Contents

Yang Guang, Thank You, Ma'am

I would have loved this book to begin with me sweeping through a rainforest lagoon, binoculars in one hand, notepad in the other; waiting, watching, creeping toward unusual animals in the midst of sexual ecstasy. Or with me halfway up a baobab tree, watching cloacas collide between birds just metres away. Or swinging on a rope above the cliff-top clefts in which female Mallorcan midwife toads wrestled and writhed in an attempt to secure access to a perky male. But no, it doesn't begin there. Instead, this book begins with a visit to Edinburgh Zoo, and features me staring at a giant panda. Or, to be precise, staring at a giant panda's bottom.

This furry backside, in the corner of her enclosure in the zoo, belongs to Tian Tian, the zoo's female panda. I want to describe this bottom a little better, but as far as mammal rear ends go it's rather nondescript. No obvious sign of buttocks, no pink genitals hanging out, moderate hairiness, uniform colour. It looks like a tasteful faux-fur cushion. I had imagined that when I saw a live panda for the first time, literary fireworks would spark. That thoughts would scribble magically onto my notepad; that I would be energised with feeling, with passion, with the awe and wonder of this most exquisite of wildlife encounters. ONLY 3,000! 3,000 LEFT! THE BRINK OF EXTINCTION! THE DESPERATE FATE OF LIFE ON EARTH! Yet ... why am I not weeping? I should be on my knees, praying for their salvation (or ours). But . . . nothing. Instead I draw a little illustration: a circle with some fur sprouting from it. That'll have to do. Underneath I scribble simply, 'panda's bottom'.

It's late 2012 and I'm paying homage to Tian Tian and Yang Guang. They are the zoo's recent acquisitions and, if all goes well, will soon be the mother and father to a new baby panda. In their first year together at the zoo it didn't really happen (I'm told by zoo staff that Yang Guang 'had trouble with her anatomy', so to speak, and that he 'missed the target'). Perhaps 2013 will be their year?

We're used to being told this, but giant pandas really are a species on the brink. Just those 3,000 or so of them remain on this Earth; a pretty ropy statistic for an animal that once roamed much of China. Victims of a loss of great swathes of their bamboo habitat, made worse by poaching and a (historical) desire for the public to see them in captivity, giant pandas deserve their reputation as a species teetering on the edge of extinction. I grant them that truth. So, I wanted to see them for myself. I've made the visit to Edinburgh, paid the entrance fee and here I stand. But there's more. For, you see, I have recently come to spend rather a great deal of time playing devil's advocate on their behalf, protecting them from an increasingly hostile public reception. In recent weeks I've found myself writing a host of ripostes, each squared firmly at those people penning comment-pieces in red-tops and broadsheets about the pandas' evolutionary dyspraxia, or their special kind of inadequacy, or their apparent inability to perform that most natural of acts: sex.

What am I to do? Sit on my hands while the pandas take a pasting? I did that for a while, but, hell, when the *Guardian* gets in on the act, you know it's gone too far. Their recent editorial (titled 'Unthinkable? Stuff the pandas') pressed the right buttons to send me into a frenzy. All the buzzwords were there: 'evolutionary mishap' (DING), 'reluctance to procreate' (DING), 'metabolically ridiculous' (DING) and the classic accusation that 'funds are disproportionately siphoned away' from other species on the knife-edge of extinction (DING! DING! DING!).

Am I the only one in the world starting to feel a bit sorry for them? These fat black and white scroungers – forever taking, taking, taking – failing too often to have the common decency to breed like the rest of life on Earth? Is it really their fault? Do they deserve such stick? No, of course not. So, I have become something of a panda apologist. For are they really sexually maladapted? Do they deserve their reputation as frigid, wasteful prudes? The more I've come to think about it . . . well, no. I'm not sure they deserve it at all, actually. Surely pandas are as sexually qualified as the rest of us, coming as they do from a long line of ancestors who managed it successfully (all the way back to our common mammalian ancestor and, in fact, back to the start of eukaryotic life)? None of their ancestors ever failed. They're the same as you and me in that respect. Pandas aren't bad at sex; they have the same batting average as every other animal (pretty much) that you've ever seen. 100 per cent. Actually, I think they're quite good at sex, based on the majority of the public's opinion of what constitutes 'good' and 'bad'.

Let me explain. Surely the fact that the male panda's high sperm count (20 times more potent than that of some bears) tells us something about their sexual nous? And surely the female's tight reproductive window, and the fact that this strategy has managed to see pandas get laid without any problems for thousands of generations, gives a hint that, until we came along, all was OK. In essence, those panda bodies know what they're doing. We're the ones that buggered their lives up. Far from being confused, sexually lobotomised bedroom bumblers, pandas are likely to be anything but, in the wild and away from us, at least.

Most of what I know about pandas comes from Henry Nicholls's book *The Way of the Panda*. From its pages I know that the story of panda sex is one of humankind imposing knowledge of HOW SEX WORKS onto pandas, expecting them to procreate, watching them fail, then looking instead at what it is that gets pandas going in the wild, and using this knowledge to make them sex-up in zoos. All sounds rather obvious now, doesn't it? Well, yes. Yes it does (for as you will see in this book, we are arguably the most sexually clueless species of the lot). The most blindingly obvious of those early discoveries about pandas is that they can smell things that we can't. Sex smells, in other words. Those panda scientists in the 20th century completely ignored the existence of such a sense, yet it seems so obvious now. To get captive pandas interested in sex, you need to let them get a good whiff of one another as the countdown to the female's fertile period begins. Specifically, the scientists found that males and females in wild panda populations keep track of one another by such communication, calibrating their sex hormones in readiness for the final act. That's why they only need a tiny reproductive window – anything else is profligate. This knowledge, second nature to us now, changed everything. Today's panda-keepers depend on such information – the smells of sex. They scoop up panda-pee whenever they can and douse wooden blocks in the stuff, liberally distributing the blocks between the separated female and male cages. That's what gets them going.

For captive panda breeding, this insight into smell and how wild pandas used it was invaluable. Unbelievably so. At one research station, where the scientists were armed with such knowledge, the captive population went from 25 pandas in 1996 to more than 70 a few years later. The lesson learned was this: if making panda babies is your trade, it pays to understand how pandas have sex in the wild, rather than sticking a few together in a pen and hoping for the best. Yet still, after all these discoveries and all this knowledge gleaned from the wild, the pandas drag around this reputation as sexually misguided nincompoops – 'evolutionary mishaps'. I find this more than a little confusing. What the hell is going on here?

I think I like pandas not for what they are, but for what they stand for – they're animals about which everyone has an opinion, but that few of us truly know. Not yet, anyway. And do you know what? I find them attractive for this reason. Yes, you can argue whether or not they're worth the millions and millions of pounds spent each year on their conservation (and I'd agree with you that it's worth discussing), but please, don't start mocking their sex lives. At least not until *you've* tried living a solitary existence in an enormous bamboo rainforest, and somehow managed to track down a mate and copulate with them on EXACTLY the right day to successfully create viable offspring. I encourage you to try it. Go out there now, Hunger Games-style. I guarantee you'll fail. And the pandas, if they could, will laugh at *you* for being so wholly useless at sex.

So let us turn now to this book. I'm rather curious about the whole 'sex life of animals' thing. Many scientists and science-writers have tackled it admirably, and done it far better than I could ever dream of doing. But, if I'm totally honest, I occasionally find myself getting a bit bored. Animal sex books can be a bit like pornography – all big breasts and whopping great penises. It's easy to get numbed to it all after a few pages. Most accounts include tales of females eating males after coitus; of sneaky males and dominant bully boys; coy hens biding their time, waiting on the sidelines as the peacocks flaunt their wares; male 'rape' of females (we'll talk later about the use of that word); infanticide; cross-dressing sunfish; male seahorses giving birth; the phenomenally long penises of barnacles; the floppy enormity that is the blue whale's todger.

There is plenty of space given to the male lion in this sort of animal sex-piece, for he is a true animal sex star; mammalian link-bait – a biological 'click here'. He can apparently go for hours at a time, and he mates up to one hundred times a day. Documentaries say things like 'Top males must possess extraordinary virility, as lionesses may require hundreds of bouts of mating to get pregnant.' But hang on . . . wait a second . . . it takes him *that many* attempts to successfully inseminate her? Weren't we just giving pandas stick for being bad at sex? On that basis, lions are hopeless! Terrible! There are reasons, of course, for the lion's behaviour: as with the pandas, it often has to do with males and females locked in evolutionary combats, or locked in combat with others of their own kind.

But I digress. In essence, all I'm saying is this: animal sex stories can occasionally play out in odd and strange ways, sometimes bordering on the pornographic. We bring too much human baggage to popular news stories about animal sex, something I have come to detest. And I've always been a bit confused about who's writing the script on stories like these. News editors? Publishers? Broadcasters? Men? Or is it just human nature to wonder about such academic questions as 'whose is biggest'? Is it human to wonder about which animal can go the longest, or which produces the most ejaculate? Is it human to want to know which female animal bites off a male's penis after sex? Are the popular animal sex stories that are commissioned just a thin veneer through which we see our own insecurities or desires, each played out in the lives of the animals we see on our TV screens? Is this all just social commentary? Perhaps. Honestly, I really don't know. But I think about it every time a female panda is slammed for needing a sperm donor or for being 'hapless' or for showing 'reluctance to procreate'. After all, pandas are just animals – interesting beyond words; mysterious residents of a planet that has brimmed with sex, without any conscious observation, for perhaps a billion years. Sex is bigger than us. And it's bigger than pandas. And lions. And barnacles.

That visit to Edinburgh got me thinking, and writing, and I wrote a piece about those pandas that got me talking with a friendly chap at Bloomsbury, the publishers. In a busy Chinese restaurant I whispered expletives across the table at him, like an international secret agent giving up military secrets. 'Am I the only one to want to know a little about the biggest *vagina* in the world?' I hissed, noodles dangling from my chin. 'Am I the only one who wonders how the Earth's movement around the sun affects how *horny* the frogs in my pond might be?' I leant in closer. 'Why do we HAVE *sex*?' I say the s-word quietly and through my teeth so no one can hear. 'Why do STICKLEBACKS have it? Why do some animals, like wasps, have *sex* and then largely DIE, and yet other animals carry on ready for more *sex* the year after? Why does a peacock's train scream 'SEX!' to us, but not the contents of a panda's pee? What's going on?'

These are the sort of questions that can seem either ridiculous, or bold and brave. I suspect that Bloomsbury thought them ridiculous, for ridiculousness also has value, if only as a yardstick against which true sense and rationality can be judged. So they went with it and I was to devote the next year of my life to sex: animal sex. I got the commission. I walked out onto the street. 'Where do I go from here?' I thought. I did a lot of thinking. I tried to approach sex with a clear mind, forgetting all that I'd learned. I started on the animals with whom we spend most of our time, the ones that are all over the place. I began investigating these, the sex lives of the everyday. The hedgehogs, the frogs, the dogs, the ducks, horses, rotifers, garden spiders. And among them I found magic. I searched further and found the fireflies, the slugs, the slug mites and the salamanders – each of which is long overdue in having its sex story told. Every one of them is a majestic sexual being, each capable of stopping us still and getting us to shut the hell up about barnacles and lions and focus our full attention their way. They show us how life on Earth truly is.

Sex got interesting again. Natural selection likes to solve a problem (as well as creating new ones), and how all animals find and have sex is one worthy of our full attention, not just a titillating nod and a wink. This book covers the others, then; the also-rans, like me, who like their sex wild, passionate, hard and fast but also, well, a bit *normal* too. For there's beauty in that sort of everyday sex. As I worked through the literature and spoke to scientists it underlined to me that, actually, the story of sex on Earth plays out not in the headliners, but in the day-to-day, the year-to-year and across and throughout the fossil record, not just with adaptive endowments and brash penis-shaped pub banter.

At this point you might have a burning question. 'But who cares?' you might cry. 'Isn't it all just sex?' Well, yes. Quite. But I suspect that knowing and understanding the generalities of animal sex lives might have some value, particularly if we are looking for a solution to how we might go about saving their greatness for our children and grandchildren. This is something I touch on later in the book: that knowledge about sex is a conservation necessity. Pandas exemplify that brilliantly (as do fen raft spiders, for that matter). For this is a great time to be alive, and it's our job to try and keep it that way. Sex allows this greatness to continue, in almost everything (except those pesky rotifers in Chapter 7). For every burly elephant seal guarding a harem, there is a hermaphroditic slug swarming over a dog turd, or a nearly extinct spider being encouraged to have sex in someone's kitchen. That's where the story is at. A panda sniffs a piece of wood; a toad safely crosses a road to find its ancestral breeding pond; a dolphin gently gooses its buddy; dogs hump. Sexual successes each and every one, each worthy of admiration, respect, greatness and further study. Each worthy of inclusion in this book.

I felt great joy while writing this book. There were moments between animals that, being a human, I would label as close to deep happiness, closeness and warmth. Against a popular backdrop of rampaging purple-headed monsters and 'o-faces' close to the demonic, there was care, tenderness and, dare I say it . . . love? (a subject that is, of course, covered in the final chapter). Whatever the story, reflected throughout its pages are those pandas, sitting there, misunderstood and ne'er celebrated for the sheer magnificence of their potential or their evolutionary history . . . if only we listened, observed and, if only now and then, metaphorically dashed our faces with their pee to remind us about the sexual perspectives of animals other than ourselves.

This is a story about Sex on Earth, and it is dedicated to Tian Tian and Yang Guang. Their pee smells wonderful . . . if you only stop and take a second to think like a panda.

CHAPTER ONE

Jurassic Pork

'The Joy of T-Rex' reads the *Mail Online* headline, followed by 'Scientist shows how dinosaurs had sex'. And I'm hooked. I click. I must read more . . .

Like millions of other readers, I'm now deeply engrossed in a pithy 'science' story about how dinosaurs did it (apparently doggy-style). 'All dinosaurs used the same basic position to mate,' says the researcher quoted in the article. 'Mounting from the rear, he put his forelimbs on her shoulders, lifting one hind limb across her back and twisting his tail under hers.' This quote is quite far down in the piece, and frankly, I'm not sure if anyone ever gets to it. They, like me, are likely to be transfixed by the pornographic collection of computer-rendered illustrations showing dinosaurs doing it.

The biggest, at the top, shows a *T. rex* male, well – how to describe? – forcing himself upon a female, from behind. She seems pinned by his weight, subdued – waiting for it to be over, while he thrusts into her rear parts. It's his face that worries me slightly. The 'scientific' illustrator seems to have given the male tyrannosaur a malevolent grin – a mildly psychotic smile is on his face, bordering on the rabid and dancing with deviance. His head is bent over hers, and he's staring demoniacally into her eyes as he thrusts. It makes me ever so slightly uncomfortable. I wasn't aware that dinosaurs could be misogynistic. She's passive and bearing the brunt, that's for sure. Her eyes are half closed; she looks accepting of her lot, her jaw slightly clenched. It's sinister.

A host of similar illustrations litter the article. Two sauropods lollop around in the water, the male reared up behind the female, forcing her tail to one side; his head is thrown back in some sort of wild rage. Is that an expression of boredom she's wearing? And why is he so *angry*? In the picture below, a *Pentaceratops* male straddles a female. He appears to have an 'o-face' traced from the back of a dodgy '80s porno magazine. These are incredibly human expressions, and I am not the first to have commented that these scientific illustrators may have been men with erections.

For an article about scientific research, it plays it loose. 'The physical challenges involved must have been formidable,' it breathlessly reports. 'The penis of a *Tyrannosaurus* is estimated to have been around 12 feet long.'

Wait . . . what? 12 feet? More than 3.5 metres? That's rather an odd statement to make when no one, ever, has found a fossil of one. Lacking a bone inside, flaccid and liquid-filled, the vast majority of animal penises are too soft, under normal conditions, to fossilise. So how do we know this one would have been 12 feet long? Statistics like these are based on little more than clumsy extrapolations from modern-day cousins, such as crocodiles, and not much else.

We know that those dinosaur descendants, the birds, possess penises (though many branches of the avian family tree have since lost them) and so did many of their reptilian cousins. Ergo, the thinking goes that, boom: male dinosaurs had penises. The 12-foot penis meme is an educated guess, nothing more. Yet here it is, given almost as fact. I sometimes wonder about things like this. Perhaps there's value in admitting more readily that we don't yet know some things? There is excitement in mystery, even more so when talking about the mysteries of dinosaur sex.

It's late December, and this popular article rattles me enough to put pen to paper. Sex isn't all about penises, after all. It's about more than just orgasms and tired porn-star illustrations. Don't get me wrong. I love imagining dinosaurian sex parts just as much as the next man or woman. And I like to go fossil-hunting. It's just that I'm not ready to throw myself into a world where we pretend we know how long a *T. rex*'s erect penis is. So I want to be clear with you about something, right from the off: this chapter is about anything other than *T. rex* penises. For fossils can tell us much, much more about sex than you might imagine. And they're old. They tell us about ancient sex, so they seem like a useful starting point for us on our journey into sex on this planet.

Fossils amaze me. They always have. Though I'm a complete amateur, it's the probabilities I'm drawn to. Every fossil-hunting trip is laden with possibility, with probability – will I be lucky today? I love it. That animals and plants can have died millions of years ago in such a way as to avoid being scavenged or properly decomposed, buried by ancient sands and muds so that hard bits (and occasionally soft bits) fossilise, and that a fraction of these fossils, after spending millions of years underground, can then be dug up to be held by primate hands, ogled by our sub-hominin eyes, manipulated and analysed by our ape machines. Thought of like this, each and every fossil is a chance event. Every single one is a five-card trick, an ode to the millions of dealt hands and wasted lives about which we will never know.

But for fossils that show us sex or sex-parts, the odds of discovery are always low. Always. A true fossil sex-searcher is therefore forced to look elsewhere. In this chapter, I hope to offer some suggestions should you be, like me, a fossil perv, or a potential fossil perv, or simply an inquisitive human being, with an interest in what makes life tick, or a fondness for making people in the pub laugh. So, let us begin. What can fossils tell us about sex, and the ancient lives of those animals that partook of it?

*

When you feel the need to immerse yourself in a world of primordial gloop, there are few places to go better than Leicester. There, among the tight streets and limping industry, you'll find New Walk Museum and Art Gallery, home to *Charnia*, one of the world's oldest fossils.

I like to visit occasionally. I like to take in the punters, sit and watch the groups of children whirling from exhibit to exhibit like flocks of starlings. Watch carefully, though, and you'll notice them zip straight past *Charnia*, which is, after all, rather boring. Try as I might, I can do little but describe this fossil as anything other than feather-like, with a pea-sized holdfast at the bottom. So primitive is its appearance that it is genuinely impossible to offer any form of anthropomorphic comment. I yearn to call it 'perky' or 'menacing' but it is neither: just a frond sticking up from an ancient sea floor. Perhaps that is why it appears to be roundly ignored by museum visitors?

But *Charnia* is ancient – 560 million years old and don't you forget it – and herein lies its charm. Famously found by a Roger Mason (then a schoolboy) in the 1950s (though it may have been previously observed by Tina Negus, who described it to a sceptical teacher at least the year before), the specimen became the first undisputed fossil found in

Pre-Cambrian rocks. It shook palaeontology. Until its discovery, people had assumed that either there was no life back then, or that it could not fossilise. *Charnia* told us otherwise. It was big news.

And there it stands in Leicester's excellent museum, behind glass and perfectly lit, like some sort of ancient scroll. Next to this display case sits something that is slightly more my style. Something you can touch. Something you can tenderly rub your face against (when no one is looking). It is a wall covered in replica fossils from a 560-million-year-old sea floor, arranged like jigsaw pieces into one enormous grey slab that has been hung on the wall. It is an ancient world, a Pre-Cambrian tapestry. Run your hands over it and you can feel the profile, the texture, of a world where the tree of animal life, as we know it, was little more than a fragile and slowly unfurling stalk.

Swirling masses of feathery strings litter the fossilised scene like galaxies, interspersed with circular lumps like enormous planets. These are either the hardened remains of rotting *Charnia*-like creatures or algal mats, or so the board tells me. At the top sits an enormous, craterous ring, the size of a pizza. Like a super-massive black hole, it looks as if it is sucking in the spaghettified remains of a galaxy. This is my favourite, largely because no one really has much of an idea exactly what it is. The circular bit is almost certainly a holdfast. But the rest? Animal? Mineral? Vegetable? Questions like these keep Pre-Cambrian scientists up at night.

The whole wall of fossils shows us creatures that lack any form of perkiness; they aren't seedy-looking, or racy, or sinister, or warm, or needy, or sorrowful. It's just . . . dare I say it . . . life looked, well, a bit boring back then?

If there's one thing that seems to link all of these creatures together it's their holdfasts, I suppose. These creatures were pioneers – global masters – when it came to holding on tight to the sea floor. Natural selection seems to have focused almost wholly on getting that holdfast exactly right, EXACTLY BLOODY PERFECT, before starting on the rest. And it's this lack of movement that I find so incredible. For these creatures, at least at this stage in life, movement just doesn't appear to have been much of a *thing*. Unimaginable to them, these investors in staying still as best they can. If they had been capable of thought, what would they have made of us, our arms, our wings, our legs? (Behold! UNSTILLNESS!).

And yet, creatures this old and this alien might have had sex. Yes, sex. Or at least we *think* they had sex. It sounds amazing. Hell, it *is* amazing. Sure, there was no thrusting or humping, but there is likely to have been sex. You know, the basics: sperm mixing with eggs, creating fertile offspring that continue (at least every few generations) to play the game. Sex.

Funisia is one such example of an early sex-fossil. It is an upright-standing worm-like animal that looks a little like a section of rope. It is revered because it is one of the first animals of which we can say with a degree of certainty that, yes, it was having sex. How can we be so sure? You might imagine digging up hundreds of the things, fossilised mid-coitus, but no – nothing like that has ever been found. Instead, we know of their sex lives because of the 'sprats'. Sprats are young *Funisia*, and in fossils they generally appear in little clusters, each containing individuals all within the same size and age band, which on the face of it doesn't sound very sexy. Except that we see exactly the same thing in modern-day animals like corals and sea-worms, which broadcast sperm and eggs out into the water, often on highly specific nights of the year. Fertilised eggs descend in the water column, hit the floor, grow and, hey presto – sprats. The products of sex. They are a little like sex's trace fossils, like a footprint or fossilised faeces. And it is one of the earliest pieces of evidence that sex was taking place in those early seas. Disappointed? Were you expecting me to offer you a fossil of two worms, end to end, enjoying their final moments? I hate to let you down, but such fossils are almost non-existent.

There are literally just a handful of fossilised animals out there that have been preserved mid-coitus. Among the most famous are those of a sprightly ancient turtle *Allaeochelys crassesculpta,* fossilised denizen of the famous Messel shale beds in Germany. Described most fully in 2012, at first glance these fossils look a little like circular nebulae rubbing together. Look a bit closer and you can almost make out a flipper, and after even closer observation, there, you can see a turtle, firmly attached to another turtle. It's rather like looking at one of those dated magic-eye pictures – stare long enough and, by God, there are two turtles having sex. It is the opposite of *coitus interruptus* – a kind of everlasting unity framed in the rocks, made only for the amusement of us and the scientists. Rightly, the fossils of these turtles captured the global media's imagination, largely because such fossilised acts in vertebrates are so rare.

Evidence of invertebrates having sex in the fossil record is only slightly more common. Invertebrate palaeontologists have discovered a total of 33 fossilised sex acts to date – many of them preserved forever in amber tombs that are relatively easy for scientists to study. New examples of ancient invertebrates having sex do turn up from time to time. In 2013, a 10-year-old boy struck gold during a show-and-tell session at Oxford University's Museum of Natural History. He brought an unknown fossil, found on holiday in Cornwall, to experts at the museum, only to be told by the scientists attending that the mysterious footprints that ran along his slab of ancient clay were likely to have been a pair of horseshoe crabs scurrying around in the throes of mating. The oldest known horseshoe crab fossils date to at least 455 million years ago, which gives another hint at the ancient origins of sex. Incredibly, another sex fossil turned up only *yesterday,* at the time of writing this chapter – two froghoppers, preserved within a hunk of Jurassic rock from China. They look to all intents and purposes as if they're cuddling – and the fossil has preserved a remarkable amount of detail.

And that, with the addition of a few interesting fossil fish (including a shark with a handlebar sticking out of its head, onto which a suspected mate is biting), is about it. With such little evidence for sex in the fossil record, can we be sure it was happening? It's an obvious question to ask, but I guess we have no reason to suspect otherwise. In his 1963 *Principles of Paleoecology,* palaeontologist Derek Ager wrote: 'After eating, the most widespread habits among modern animals are those concerned with sex, and there is no reason to suppose that this did not raise its allegedly ugly head millions of years before Freud.' It's a useful quote, and I'd add only this: if animal life can be viewed as a family tree, sex appears almost universally on each outer twig: we only have to look around us at the sex-obsessed animals alive today. They're everywhere. Sex must have been there on the earliest branches or even at the root of the tree itself for it to be so universally commonplace. It must have been there, but the exact details of it remain hidden behind the curtain of time. Hidden forever, perhaps? Encouragingly, no. For there are other ways to learn about ancient sex . . .

It was at this point, in the winter months early in my journey into animal sex, that I thought it would be a good idea to spend a few hours wandering the meandering paths of the famous dinosaur park in London's Crystal Palace. And I'm glad I went.

Only a couple of minutes' walk from Crystal Palace station it lies, a magnificent place. It's like a Victorian Jurassic Park; the sort of place you go to once in your life, and if you've never been I implore you to make the trip. It's a little like a crazy golf course; it has nicely kept paths, well-pruned trees, well-weeded and ordered flowerbeds. There is also a crazy-golf-like tweeness, a novelty in the characters that stand beside you, watching you cartoon-like from the sidelines. *Megalosaurus*, *Ichthyosaurus*, *Iguanodon*, *Mosasaurus*. From the entrance to the park, you can see their white chiselled forms poking out from the tree-tops. You can even make out some of their expressions: distrustful, brutal, rugged, wild, pissed off; the best expressions that Victorian brains could imagine based on the meagre evidence that was available to them through fossils of the time.

As with Leicester's New Walk Museum and its Charnian vistas, the Crystal Palace dinosaurs sit in a special place in my heart. A portal to my childhood. When I first came here as an 11-year-old, it was with my first camera (obtained by returning tokens from Mighty White bread). The square photos in that, my first photo album, show me and my family posing in front of each and every one of these mighty beasts. My dad with his loafy side-parting, my mum (those are very big earrings), my brother (double-denim) and sister (a face that says 'I'm too old for this') and me, the youngest. In each picture I'm posing in front of one of the white monsters, like a model in a bikini sprawled over the bonnet of a cheap sports car. In those 1980s photos you can see how decayed the dinosaurs had become (they've since been done up). Their pale, lichen-covered faces peek out from overgrown ivy and brambles, like grotesques from a fallen church. But my face glows in each one. I was in my Peak Dinosaur Phase.

What's behind this dinosaur phase we see so often in children? It's always confused me. Popular thinking has it that young kids love dinosaurs for a variety of reasons, all (as far as I know) untested. Perhaps it's that dinosaurs show all the fearsome scary attributes of a monster, but are viewable through the safety goggles of ancient time? Perhaps it's because dinosaurs are an area of knowledge about which most grown-ups understand very little and in which young people can wield some early authority over their parents, siblings and peers? I have recently heard a third theory – that

somehow dinosaurs represent adult authority, and that
children are drawn to them through a complex mix of
parental admiration, desire for security and safety, and a love
of the dominant alphas. Why did I love these enormous clay
models so much? For me it was something about being near
them, or near drawings or books of them – gathering
perspective, imagining; making monsters real. I would
never have my encounter with a real-life UFO but maybe,
just maybe, I would be the one to dig up a new fossil and
name myself a new dinosaur. With dinosaurs, all of us had
(and still have) the potential to be pioneers, as long as we
have tools for digging and brains for thinking (hell, that
thought excites me even now).

Now restored, the dinosaurs at Crystal Palace park have
gathered a kind of cult following. This is probably because
there is a naiveté to them – being, as we now know they are,
largely and hilariously wrong in all manners of posture, size
and overall demeanour, but once, to the Victorians, right.
Or, at least, as right as could be imagined with the evidence
to hand. As funny as it sounds this really was once a state-
of-the-art place – a Victorian exhibition that showed the
world just how much scientists were discovering about a
world previous to the one that humans had inherited.
Perhaps for the first time, science was having its day – telling
stories to the public based on truth, as magical as anything
that religion or the myth-makers could dream up. The park
was largely the work of Richard Owen, he of dinosaur-
naming fame. His 'Dinosaur Court' opened in 1854, made
real by the modelling skills of Benjamin Waterhouse
Hawkins, a Victorian with the hands of Michelangelo and
the entrepreneurial spirit (and possibly ego) of Richard
Branson.

Lauded though these two men were at the time, the clay
models eventually ended up as testaments not to them but to
Charles Darwin, fitting as they did within his celebrated
scheme rather than Owen's. When *On the Origin of Species*

was released later that decade, Darwin's theory managed to explain *where* the dinosaurs may have come from, and why they appeared at all. It was simple. Natural selection, with time on its hands, was largely responsible. In Darwin's world these creatures – their broad shoulders, long teeth and monstrous visages – were chiselled by unthinking hands, through a selection of survivors ('a series of successful mistakes', as the geneticist Steve Jones succinctly puts it). They weren't chosen by God. They weren't chosen by anything. They just popped up, had their time and then wobbled off this mortal coil through no fault of their own, leaving the rest of us to adapt and start a new age and a new chapter, The Age of Mammals (a little part of the bigger story, The Age of the Nematode Worms).

Walking among these enormous Victorian monsters on your own is rather wonderful. I enjoyed my hours there, weaving in and out of the dappled sunlight, left and right past the baby-buggies and the joggers; stopping and standing and wondering and chuckling at the simplicity of Owen and Hawkins's creations. If you should visit, and you are of the 'I'm just here for dinosaur sex, officer' ilk, you might be disappointed. None of them are chiselled in any form of union, which is really no surprise; these were Victorians, and even the sight of a baboon's bottom was enough to make them go weak at the knees. So here, in this park, each gargantuan reptile has a groin as smooth as a billiard ball. No swollen cloacas. No penises. No o-faces. Nada. These Victorian constructions are wholly sexless dinosaurs, without obvious male or female features.

But there is a clutch of creatures within the park that allows an observer to work out who's who, so to speak. And they aren't dinosaurs. Head across the park, back and around the lake and toward the Irish elk, for here there is a whiff of sex. Sexual dimorphism, at least. Walk over toward the shoreline of the big lake and you'll see them. One statue stands proudly on a boulder, its antlers protruding high off

its head like an elaborate fascinator. So big is its skull
furniture that it looks as if it's wearing a Russian satellite. Its
muscular body is riddled with slug trails, but the pink and
yellow lichens add some richness to its faded form. Keep
looking, though. There, next to this obvious male, a female
is lying down. She has no antlers, and slightly effeminate
eyes like a Disney character. A little baby elk, too, stands
between them. It is too young to tell which sex it'll be –
such can be the way with sexual dimorphism. They are
cracking good beasts, and in their bones and armaments we
can see sex.

The Irish elk was neither Irish nor an elk but rather a
deer, a bloody big deer. The biggest of all, in fact – more
than two metres (6ft 6in) at the shoulder, and with antlers
that came in at 12 feet from tip to tip (or, if you prefer, one
T. rex Penis Unit). The Irish elk is thought to have died out
a shade under 8,000 years ago, for reasons that remain largely
a matter of guesswork. The male in Crystal Palace stands
proud, a totem to a theory that remains largely unchanged
since Darwin's time, and is without doubt his second biggest
postulation – his theory of sexual selection.

After the publication of *On the Origin of Species*, Darwin's
all-encompassing theory of natural selection fast became
popularised as a fight to the death between predators, prey
and competitors ('nature, red in tooth and claw', as Tennyson
famously put it). Yet, somehow, all did not seem right to
Darwin. Antlers, for one thing, didn't seem to stack up. The
general consensus at the time was that antlers are, to all
intents and purposes, formidable weapons. To most people,
they fitted right into Darwin's theory. But Darwin couldn't
shake the fact that they were big – too big. To him, the
whole thing seemed so *expensive* (antlers are physically costly
to grow and are, after all, shed each year). There seemed too
many drawbacks for Darwin. Too much waste. It troubled
him. It was time to revisit the drawing board.

In 1871, he published *The Descent of Man, and Selection in*

Relation to Sex, in which he referenced adaptations such as stupendously big antlers and came up with another theory that might explain what was going on. Though it proved less of a lasting hit than *On the Origin of Species*, this later book toyed with the idea that horns, rather than being for killing predators and shooing off conspecifics, might be a useful way for males (more often) to 'display' their 'quality' (fitness) to females. 'If, then, the horns, like the splendid accoutrements of the knights of old, add to the noble appearance of stags and antelopes, they may have been partly modified for this purpose,' he wrote.

Darwin was onto something. He was arguably the first to see that females could drive the evolution of male traits, through choice. His theory of sexual selection was born, a form of evolution capable of skewing behaviours and body shapes toward wild exaggeration, driven by reproductive success, rather than simply survival. Preposterous antlers, bowers, horns, manes, jaws, tusks – that sort of thing. Not exclusive to males or females, but often a process capable of leading to the absurd. And by 'absurd', I mean, of course, peacocks.

It is said that tigers stroll up to male peacocks, pin them down with a big paw, and quietly snack on them like cocktail sausages. Theirs is the evolutionary equivalent of being gifted with a very gaudy jewel-encrusted gammy leg. Useless for evading capture but, cor blimey, look how it shines in the sun. Only sexual selection can produce magic such as this. Darwin began viewing females of some species, particularly elks and peacocks, as drivers of evolutionary change, not as prissy extras that stood around gawping at males in battle. This was indeed revolutionary. It was a female power of sorts (though it can occur in either sex, or both) and Darwin was first to name it. Though the theory isn't perfect – scientists continue to argue about the exact mechanics behind some aspects of sexual selection – it's held up relatively well since Darwin's day. And those mighty antlers that stand like giant thunderbolts above the Irish elk are part of the story. Visit that park in Crystal Palace. Pay homage.

The idea of sexual selection acting on males, equipping them with the means to fight off rivals to acquire a female, was broadly accepted at the time. It accounts for a great deal of adaptive nastiness between individuals of the same sex (often males): burly elephant seals, stag beetles, stalk–eyed flies, goat flies, snakes (including the adder), ibex, antelopes, western gorillas, and even giraffes. All of them may have evolved traits to battle against, or at least intimidate, other males. People in Darwin's time bought into it. But females acting like pigeon-fanciers? Picking traits they 'liked', selecting the males that 'impressed' them most? The idea was comical to many, and this part of the theory failed to gain traction. How could a female peacock 'choose' a mate? How could a creature with a brain as tiny as a peacock's possibly 'like' or 'prefer' a male? It couldn't, surely? Alfred Russel Wallace (co-founder of the theory of evolution by natural selection) famously had cold feet about the whole thing. He thought it ludicrous, and instead favoured the notion that males were often more colourful and loud, not to woo a female, but because of their 'superabundant energy' during the breeding season (yes, you read that right). Needless to say it took a long time for the notion of female choice (now 'mate choice') to gain traction among biologists. According to the evolutionary biologist Michael Ryan in a recent *National Geographic* article, the idea was still scoffed at during the 1960s and 1970s.

How wrong they were to scoff. There is now a healthy and well-researched bevy of creatures known to demonstrate 'mate choice'. They include swordtail fish, hens, crickets, mice, bowerbirds, guppies, wolf spiders, albatrosses, weaver-birds, lions, goshawks, widowbirds, auklets, manakins and many, many more. There are likely to be thousands and thousands of other such examples, so many that scientists have barely scratched the surface.

'The sexual struggle is of two kinds,' Darwin wrote in *The Descent of Man*. '. . . in the one it is between individuals of the same sex, generally the males, in order to drive away or kill their rivals . . . whilst in the other, the struggle is likewise between the individuals of the same sex, in order to excite or charm those of the opposite sex.' Sex can buy you flowers or flame-throwers; sexual selection decides on which. In elephant seals, where competition on the breeding beach is high, it selects for big bulky marine marauders. In dense forests, it works on the voices of Japanese bush warblers and purple-crowned fairy wrens. In parrots, it works on the feathers. In stag beetles, on the horns. In giraffes, it might be working on the necks (long necks being all the better for thwacking an opponent to the ground). It matters. Sex really matters. And that's because survival without sex leads nowhere in evolutionary terms. Survival *with* sex is the only game in town for almost everything.

And so we must stop, take breath, for a second at least. For there was a reason I was drawn toward fossils for the first chapter of this book. It wasn't only because I started writing in winter, when sex is, for the most part, in preparation. I was drawn to fossils largely because it seems to me that, through them, we can start to understand a key principle. Sex isn't just a thing that animals do. Sex has acted on us, on our animal shapes and behaviours, for millions of years. Too often we consider it a physical act, but actually it's more than that. Sex is written on bodies: on morphologies, in the bones, on the hard bits that fossilise. Sure, we aren't likely to find that elusive fossil boner any time soon, but there are whole fossil-beds filled with sexual creatures waiting for an analytical eyeing-up, or thoughtful consideration by someone who asks the right questions.

This is one of many new frontiers in sex science. And it is through science writers like Dave Hone (who blogs at the *Guardian*), Darren Naish (*Scientific American*) and Brian Switek (*National Geographic*) that such stories are beginning to be told. This lot, and their peers, look through dinosaur books and journals and see sex everywhere. They don't just look at a flying *Pteranodon* and wonder what its aerodynamics were like, they look at the big crest sticking out of its head like a witch's hat and wonder if it's a male or a female thing. They don't argue about whether or not *Stegosaurus* used those big plates along its back for protection, they marvel instead about whether they were used for communicating messages that revolve around sex. *Triceratops*'s big neck frill? 'My, that'd make an excellent advertising hoarding,' they think. And what of the big honking crests that belonged to the hadrosaurs? Were they, literally, for honking? Yes – there is a wonderful fossil of a *Lambeosaurus* that shows the head crest in enough detail to determine its acoustic properties. How many of these ancient dinosaurs were singers like their descendants, the birds? Perhaps many of them. This is a fascinating new world of information, and there is not a whiff of penis-size anywhere. These are sex stories that hypothesise – that ask big sex questions – about everything else, from feathers to crests to battle-hardy craniums and startling body art: the full gamut that evolution bestows on sexual creatures, in other words. This is likely to be an exciting decade.

So we return to *Tyrannosaurus,* the *cause célèbre* of Cretaceous carnage. We may never know the measurements of their genitals (male or female), but we can ascertain some details of their sex lives nonetheless. Allow me a moment to discuss them.

Although only 50 or so fossil specimens of *T. rex* have been studied (spanning about two million years), it's the species at which, arguably, most speculation has been thrown. So famous is *T. rex* that, in fact, it's hard to imagine a world without it. First discovered in 1892 by the famous palaeontologist Edward Drinker Cope, it was officially christened 'tyrant lizard king' in 1905 (too late to feature in Owen's Dinosaur Park, alas). Since then *T. rex*'s famous skull has appeared, like a *memento mori*, in references to dinosaurs in popular culture across the generations. Like Kate Middleton on the front page of a red-top, you only have to mention the *T*-word in popular culture and the punters will come running (as the dino-penis story demonstrates nicely). And so let us turn to some simple facts about *Tyrannosaurus*. They are these: 12 metres (40ft) long, four metres (13ft) high. Head the size of a cot. Arms the length of a baby. Teeth as long as a baby's arms. Basically, among the biggest, most ferocious land carnivores the world has ever seen – each fossil a cherished find, and one that has drawn the attention of many a scientist (and that's an understatement).

Through the 20th century, thanks to these scientists, thoughts on how it might have moved around and behaved have changed. Incredibly so. Famously, where once we thought of *T. rex* strutting awkwardly, a bit like a man in a Godzilla costume dragging his tail through Tokyo's streets, we now consider it a smooth, agile mover with tail held aloft. Where once we imagined it feeding on dead things, most scientists now consider *T. rex* to have been both predator and scavenger – much like a massive modern-day hyena. It really has been the subject of incredible scrutiny. Type '*Tyrannosaurus*' into Google Scholar and something like 13,000 results come up. Type in any other dinosaur? You'll be lucky to get 2,000. *T. rex* is a phenomenon, yet we are only now getting to grips with its sex life.

Only recently, for instance, have we established a fool-proof way of telling the male fossils from the females. It happened in 2000, after Jack Horner and his team discovered not one but five *T. rex* fossil skeletons. One of these, as is often the way, was stuck 20 feet up a cliff face. 'B. rex', as it was named, proved thoroughly awkward to get out of the rock. So heavy was it for the hired helicopter to carry that in the end it was decided that a leg bone would have to be broken in two for transport. This was a key moment in the story of how to sex a dinosaur, for it allowed Mary Schweitzer

– an expert in microscopic study of dinosaur soft tissues –
access to the good stuff found inside the fossil bones of
dinosaurs. When analysing fragments of the fossil bone
tissue inside this leg bone, she found something unusual: an
odd-looking layer of tissue called medullary bone. Medullary
bone was a big deal, since it was also known from birds. In
female birds, an aggregation of such material appears in the
shafts of the long bones in their hind limbs. It acts like a
calcium-rich storage area for the material required for
making eggshells. It's a lady thing, in other words. So B. rex
was a female tyrannosaur, the first known. Science had a
new way to find out whether the best quality fossil dinosaur
skeletons were male or female (or at least females in the
process of producing eggs).

This new technique for sexing dinosaurs could now be
paired with other life-history details gleaned from other
dinosaur fossils. By examining growth rings in these fossils
(which work a little like trees) and looking for the medullary
bone formation, scientists could then work out at what age
dinosaurs started having sex. This simple approach worked
wonders in terms of our knowledge of the sex lives of the
dinosaurs. The result? 'Dinosaurs lived fast and died young,'
as Brian Switek puts it in his excellent *My Beloved Brontosaurus*.
Tenontosaurus, *Allosaurus*, *Tyrannosaurus*: all early mums.

It may have taken a long time, but at last we can see sex
written in (at least some of) the bones of ancient life. And
we have B. rex and that imposing cliff face to thank for it, as
well as the keen eyes and impressive microscopy skills of
palaeontologists like Mary Schweitzer.

Not all of *T. rex*'s features may require such technical
wizardry or specialist analysis to uncover their hidden sex
lives. What of those pathetic arms? Could they have had
something to do with sex? They're certainly peculiar. Even
the biggest, barely longer than a toddler (to continue the
theme of defenceless babies and hungry dinosaurs), they are
tiny. Ridiculous, really. Could they have been a product of

Darwin's theory of sexual selection? It sounds laughable, but the idea has been put forward seriously by scientists in the past. Henry Osborn, official name-giver to *Tyrannosaurus rex* in 1905, guessed that they could have 'served some function, possibly that of a grasping organ in copulation'. Yet others argued, perhaps more persuasively, that they are nothing more than useless vestigial limbs, signs of ancient ancestral history (much like our tiny primate tail, the coccyx).

But another argument is surfacing about those tiny arms, and it brings *T. rex* back into the realms of sex. Could it be that these weak limbs were for display? Some scientists are drawn to the idea of *Tyrannosaurus* using them to strut, much as an ostrich might use its wings today, fanning and flapping around a potential mate during courtship. This might explain why these bones often appear fractured or damaged. And they might have been feathered, too, of course – one of palaeontology's most recent and fractious debates. For almost a century, artistic representations of *Tyrannosaurus* had them scaly and dry, but recent fossil finds from China show that at least some of their close relatives were covered in a 'fuzz' of downy feathers, something argued by some to have been used for display (or later modified through sexual selection for display purposes). Indeed, at least some dinosaurs may have been much more colourful than we could ever imagine. Something sexual selection might, at least sometimes, have taken advantage of.

Since 2010, light is being shed on fossil colours like never before – mainly through the pioneering technique of searching for melanosome structures within the surfaces of fossilised feathers. Melanosomes are microscopic packets of pigments, each with its own shape and size that correspond to the colour it gives off. By looking at the frequency of these melanosomes in well-preserved fossils, scientists are now slowly reimagining the colour palette of ancient creatures. The result? Some ancient birds had iridescent sheens, for instance, and some had monochromatic stripes.

A thorough going-over of the melanosomes on a particularly well-preserved *Anchiornis* specimen (a crow-sized four-winged dinosaur, which exhibits a number of traits seen in modern birds) reveals long white flight-feathers with black tips, and a plume of ginger feathers sticking out from the top of the head, a bit like a magpie with a straw hat on its head. Sounds sexy? Yes. That's because it probably was.

So maybe *Tyrannosaurus* was similarly well attired? Maybe. It's too early to tell, but its family history suggests that feathers played a part and had at least some selective advantage. 'We have as much evidence that *T. rex* was feathered, at least during some stage of its life, as we do that australopithecines like Lucy had hair' is how Mark Norell of the American Museum of Natural History has put it. It's surely only a matter of time before we repaint Jurassic Park's fiercest resident in colours that stink of sex. And it'll be fossils (none of them of penises) that tell us what's what.

And so we must return there, back to those penises. I apologise for bringing this up again, but you will remember that I began this chapter complaining about those 12-foot dinosaur penises. I tried to paint you a word-picture of the popular representations of dinosaur sex – panting, yelping, screaming 'o-faces' – but sex is about so much more than just this, and I hope I have managed to go some way towards outlining the myriad ways it affects animals and their evolution.

It shapes us; it has shaped all of us, probably every part of us in some small way. It's in our bones, our behaviours, our feathers and our faces, whether we like it or not. We are all touched by sex. Yet it's only recently – within the last 50 years or so – that we have begun to understand just how much of an impact sex has had on our lives, and the history of life on Earth, the history of all things. It's only in the past few decades that Darwin's ideas on sex have been revisited and taken more seriously, and I find that exciting. We're pioneers still, even in this scientific age. We are living a kind of scientific sexual revolution; now, today, 150 years or so after the publication of that benchmark *The Descent of Man*. And about time too, you might say.

I enjoyed wandering around Crystal Palace park while researching this chapter. Strolling among those big outdated dinosaurs and plesiosaurs and mosasaurs and standing in the shadow of the flamboyant headwear atop the ancient Irish Elk's head. It's a place to truly understand the notion of knowledge, to realise that our best bets are based on evidence, on fossils, on scientists looking at stuff and working out their function. Go there, visit that park, sit on a bench and think about sex. I implore you.

It excites me to think of the dinosaur books that our children will devour; that they might contain illustrations of dinosaurs dancing, wrestling, dazzling conspecifics with their ornaments or battling their brethren with armaments. All Darwin's. All Earth's. A story almost as old as time itself, that goes on all around us, but that's only now being unlocked. What a place, then, to begin a journey like this.

CHAPTER TWO

The Irascible Hulk

'They ate blubber cooked with blubber, had blubber lamps. Their clothes and gear were soaked with blubber and the soot blackened them, their sleeping bags, cookers, walls and roof, choked their throats and inflamed their eyes'.

Alas, poor Levick. George Murray Levick: polar explorer, abandoned for months in the back of beyond more than a century ago. While Robert Falcon Scott's team scoured the southern regions of Antarctica (as part of the 1911–12 *Terra Nova* expedition), Levick was part of a six-man 'northern group' whose job it would be to discover more about the region's frozen interior. In that aim, they had difficulties. Unable to navigate a safe route inland, they found that they couldn't get back to the *Terra Nova* either, it being unable to get through the pack-ice to rescue them. They were stuck. There was nothing for them to do but hole-up (literally – they found an ice-cave in a place called Inexpressible Island) and wait it out until spring, when the sea ice would melt and the *Terra Nova* could return to get them. Blubber was pretty much all they had. That and time on their hands.

Levick chose to spend much of that time observing penguins. He was to become a world authority on them (what else was there for him to do but look at penguins?) and he would later publish a book, *Antarctic Penguins*. Successful though the book was, there were some things he could never have included. For there, amid the drama he noted – the closeness, the monogamy, the *March of the Penguins*-style tenderness – he also witnessed a litany of perversion. A really striking list of sexual behaviours, displayed by the Adélie penguins in particular. The list included sexual coercion, sexual abuse of juveniles, 'murder' and necrophilia. At one point he saw male Adélies mating with dead females whose bodies had been left cold and lifeless for over a year. In his notes he scribbled the worst bits down in Greek, for fear that a casual reader would stumble upon them and suspect him of having some sort of psychotic episode in those bleak southern wilds.

Though the information was kept out of his other published works, Levick was scientifically savvy enough to write up his observations privately. They became a part of a secret paper, handily titled 'Sexual Habits of the Adélie Penguin'. This was clandestinely circulated among a handful of experts, like a dirty novel in a school playground. Scurried away. Hidden. And then, mysteriously, it disappeared. Forgotten. Lost. The paper might have vanished forever had not Douglas Russell, curator of birds at the Natural History Museum (Tring) rediscovered it in 2012, among records and scrawlings from Scott's expeditions.

Among his scribblings (re-accounted in 2012 in the journal *Polar Record*) Levick describes '. . . little hooligan bands of half a dozen or more ... [that] hang about the outskirts of the knolls, whose inhabitants they annoy by their constant acts of depravity'. According to Levick, nothing seemed to deter the lust of these male Adélie penguins; they tried it on even with injured females, corpses or chicks (one was reportedly 'misused before the very eyes of its parents'). Some males, in the absence of dead penguins or chicks, simply chose instead to have sex *with the ground* (even bringing themselves to ejaculation). One can imagine Greek symbols littering the pages of Levick's notebook (what is Greek for 'THE COLD HARD FLOOR'?).

Now, I bring this up not to relish in the details (though there is a bit of that), but rather to highlight that Levick – an academic, a scholar and a respected man – baulked at the idea of talking publicly about the sexual observations he had made of those Adélie penguins. He wasn't ready to talk about it. Academia wasn't ready to talk about it. Society wasn't ready to talk about it. The science of sex ... well, it just wasn't a *respectable* thing. If the paper tells us anything, it's of the mental and moral struggle that scientists had when trying to ask scientific questions about the sex lives of animals.

Other academics fought, like Levick, with the notion of writing graphically and scientifically about sex and sex parts. In *The Descent of Man, and Selection in Relation to Sex,* Darwin apparently battled with his publisher (and won) to include the S-word in the title, though like many of his time he chose to write the more candid bits in Latin; his daughter Henrietta later acted as his editor, and was apparently fond of the red pen for his more graphic passages. For those that could read it, Darwin's comments on monkeys' bottoms (which he knew were greatly enlarged genitalia) were derided by at least one social commentator of the time.

Carl Linnaeus (the self-styled 'King of Botanists') also had rather a flair for Latin blue-talk, and this too tried the patience of his peers. He used the words 'vulva', 'labia', 'anus' and 'pubes' when describing the anatomy of one clam species, and he famously compared the sepals of some flowers to the outer layers of skin of the vulva, and the petals to the inner layers. The result? Raised eyebrows and side glances from many in the field. There was vocal opposition, too. One rival accused Linnaeus of 'loathsome harlotry' by using such language, and Goethe allegedly expressed concern at young people and ladies being exposed to Linnaeus's gross 'dogma of sexuality'.

These were long, hard centuries, tough for sex to gain traction as a subject worthy of scientific study. You were essentially hindered unless you knew the right people, were part of the right gentlemen's club and could read Latin or Greek fluently. Sex was the biological equivalent of dark matter. There, but ... well, theoretical. Untouchable, somehow. Untestable.

Though it seems ridiculous to us now, this attitude continued right on into the 20th century. Far too long, really. But then things changed, and in a small way this was down to one single species; a lowly animal propelled forward as a paragon of role reversal, a sexual celebrity, and a mascot of playful youth. And all you need to see it for yourself is a jam-jar. That and an appreciation of the intense haranguing rage that redness can bring on ... I am talking, of course, about the three-spined stickleback, the focal animal of this chapter.

*

'Come in, come in.' My guide, Iain, leads me into a small, high-ceilinged room that hums with strip lights and water-filtration units. 'Welcome. Please, come in.' He ushers me forward, smiling. I step into a room full of shelves. Tank after tank after tank of sticklebacks look back at me. Some big, most small – some seem little more than nail-clippings. They move like little amorphous clouds from one edge of their tank to the other. The air in the lab is damp and humid, and the sound of water dripping through pipes makes me feel as if we're miles below the surface of the Earth. I need a wee, too.

'In most of these tanks are our babies – our recent offspring,' says Iain, allowing me time to wander around from shelf to shelf, like I'm buying a pair of shoes. I stare at the first tank, at the stirring mass of tiny fish, each the size of a staple, with a pair of eyes like full stops at the front end. They mass slowly from one side of the roomy tank to the other, then suddenly streak nervously to the corner as my enormous face peers in. 'And who might these be?' I ask. Iain checks an annotated sticker on the side of the tank, like a doctor pulling out notes from the end of a patient's bed. 'Hatched three weeks ago, these ones were.'

I continue to peer curiously into their world, as Iain makes his way to the next set of tanks. 'Some of these ones are a bit older – these ones four weeks, those ones over there are five weeks.' There must be about 10 or 12 tanks along this wall alone. 'And what are they all for?' I ask. 'These are the stocks that we'll use in the coming year – we typically use wild-caught fish as parents, but then many of the behavioural experiments we undertake are on these fish, the ones that we know everything about, like their growth-history and their developmental history.' These are the second-generation test subjects, then, companions in our grand pursuit of knowledge about how stickleback sex works, and the evolutionary factors pulling the strings behind the scenes. This is just one of the modern faces of sex science.

In a larger tank at the end swim some 20 adult sticklebacks, each about as long as a business card. Iain points out a couple that have a blue tinge to the upper side of their eyeballs. 'And those are the males – that's what they look like as they gain their breeding coloration.'

You will probably remember what a stickleback looks like in its spring finery, so common is the image in popular representations of nature. Big, bright red belly and blue dinner-plate eyes. A small cowlick of spines running along the top of the body. And that's about it, to most people at least. It's become nature's ClipArt. A logo of childhood encounters with nature, or so we're told.

Most of the sticklebacks I've inadvertently caught while pond-dipping are non-breeders, immature males and females or those outside the breeding season. Silver, darting and rather nondescript. Outside the breeding season they school together in ponds, lakes and rivers, picking at water fleas, insect larvae and waterlice – males near the bottom, females generally near the surface. Sometimes the sexes interact, but only to peck at scraps or to attempt to steal morsels from one another. They're scrappy, pestering little fish, really (Iain tells me that in fisheries they're referred to as 'trash fish'). One minute they're picking at the leaves of curly pondweed, the next they're tugging the legs off a poor water hog louse. Famously, though, in late spring all hell breaks loose in the world of the stickleback. A blistering rouge spreads from under the chins of the males and across their flanks. Their eyes become sky-blue and their waking hours are filled with an urgent need to engineer things. The male becomes edgy, narky – boiling with rage and fury, like an over-caffeinated logistics manager, surrounded by imbeciles, underlings and other annoyances. Sticklebacks are The Hulk of freshwater fish. And no one likes them when they're angry. Except the sex scientists.

The stickleback is a fish I wanted to see early in the process of researching this book, and I'm delighted that Iain has invited me over. In return for nothing more than the thrill of talking about his three-spined subjects, Dr Iain Barber of the University of Leicester has taken me under his wing for the morning. A passionate advocate for stickleback-kind and a global expert in fish sex and its evolution, he's everything George Levick could have become, had he lived just a century later. 'Please, feel free . . .' he smiles, and gently opens his hand, beckoning me to investigate the tanks on the far wall. He checks at dials and notes, while I stand and silently move from tank to tank, eyeballing each and every resident. It's calming. The churning filtration unit behind us is the only sound. Peaceful. Like being in a screensaver.

'A man can sit for hours before an aquarium and stare into it as into the flames of an open fire or the rushing waters of a torrent,' wrote the great Konrad Lorenz. 'All conscious thought is happily lost in this state of apparent vacancy, and yet, in these hours of idleness, one learns essential truths about the macrocosm and the microcosm.' Lorenz was part of the 1973 Nobel Prize-winning team (alongside Niko Tinbergen and Karl von Frisch) that founded a special branch of animal behavioural study, otherwise known as ethology. And Lorenz had a special place in his heart (and his lab) for sticklebacks.

This trio of academics was something special. Rather than just putting rats and pigeons in boxes and getting them to tweak little levers, their approach was to get out there and watch animals in the wild, or in labs built to mimic the wild as closely as possible. They encouraged other scientists and academics and (importantly) citizens to ask questions about cognition, sociality, and the physiological and molecular bases for animal behaviours. They added scientific foundations for what is now a key pillar of zoological understanding. They encouraged scientists to study life-history, and the adaptiveness of given behaviours, including those relating to sex. They talked about sex scientifically, and the scales started to fall from the prudish eyes of academia. And there, right at the start, were those sticklebacks. Then, as now, being watched and observed in tanks like these . . .

One famous aspect of Tinbergen's research on sticklebacks focused on something termed 'supernormal stimuli', a phenomenon familiar to anyone who studies psychology at school. By constructing wooden models of sticklebacks with undersides of differing shades of reds, Tinbergen showed that sticklebacks grew steadily more aggressive the redder the fake 'fish' inserted into its realm. He constructed sticklebacks redder than anything nature could produce, and watched (perhaps delighted) at the fury that his models elicited in the males. He created supernormal wooden models of fish that were redder than the most wonderful sunset, and observed the resulting catatonic reactions of male sticklebacks unused to such a sight in nature. Perhaps no human eyes had ever seen such rage?

Konrad Lorenz's interest in sticklebacks was related to his work on animal aggression. And sexed-up sticklebacks provided some of his most famous observations. 'The fighting inclinations of a stickleback, at any given moment, are in direct proportion to his proximity to his nest,' Lorenz writes in *King Solomon's Ring*. 'At the nest itself, he is a raging fury and with a fine contempt of death will recklessly ram the strongest opponent, or even the human hand. The further he strays from his headquarters in the course of his swimming, the more his courage wanes.'

Easy to catch. Easy to observe. Easy to breed. Easy to manipulate. The stickleback had found its way into sex science's heart. It's a model animal ('a super-model', Iain calls it).

Back downstairs, Iain leads me around to the other side of his stickleback lab to a much longer tank. Like the others, it is as clean as a whistle. A layer of small pebbles lies on the bottom and a transparent tube bubbles in the corner. This tank is different, though. It appears strangely devoid of residents. To the left- and right-hand sides pondweeds protrude from the pebbles, creating a kind of stage area in the middle. And there, on the bottom in the centre of the tank, lies a host of messy black threads, draped untidily across the gravel. They look rather like flowers thrown onto the stage at the end of a performance. 'This is one of the tanks where we undertake our behavioural observations,' Iain tells me, before pointing to the threads. 'And this used to be a stickleback nest,' he says matter-of-factly. Ah, OK. I should have known this. 'The male, once he's raised his young, pulls the nest apart and starts on a new one – and those . . .' He points behind us, smiling. 'Those are the offspring from that same nest; the baby fish that this year will become our study fish.'

It's too early in the year to see sex for real now, so Iain suggests I might like to watch some of his fish-sex videos upstairs in his office, and I readily agree. This sounds, as I write these words, a bit weird, but actually, no – no wait, it *is* a bit weird. But this is science, and Iain is a respected scientist. That makes it OK. Plus there is a distinct irony in my being prudish about sex, while writing about Lorenz and Tinbergen and mocking Levick for being such a prig.

Iain and I head upstairs. After numerous staircases and corridors, we arrive at his office and he sits me down at his desk, where he prepares a sex demonstration video on his desktop monitor for us both to watch. There is a bit of silence at this point and I sit there, in his chair, feeling like a student once more. It dawns on me that, even though I've tried many times over the years, and it's something naturalists seem to always know lots about, I've never actually *seen* stickleback sex other than in textbook diagrams. Iain's computer speakers crackle, and he clicks on an icon to show the movie in full screen. A title scrolls up. '*Testing the buoyancy of nesting material,*' it says. The screen fades from black and reveals an enormous, gorgeous, bright red male stickleback, hovering purposefully in what looks like the same tank that I've just seen downstairs. He has in his mouth a bit of black thread and he swims upwards, before dropping it and watching carefully what happens. The thread slowly falls through the water, and he quickly grabs it again. He swims higher this time, and drops the thread once more, desperate to understand as best he can the metaphysical properties of this tiny strand, something you or I would probably floss our teeth with. He picks up another and repeats the process. Then another. Then another. Each time the red-bellied stickleback watches intently as the thread floats quietly downwards. He tries this with some other bits of thread, assessing invisible qualities in each and every one.

Then, suddenly, he finds it. The perfect thread. It floats down in exactly the right way. Not too heavy. Not too buoyant. Perfect. Though this one looks exactly like all the others, he plugs it confidently into the gravel before thrusting his nether regions against it, sliding himself across the thread whilst squirting his spriggin (a sticky substance that he produces from his cloaca – the technical term for the genital opening in amphibians, reptiles and birds, one that also serves as the exit-point for the urinary system and gut) to glue it firmly in place. 'He's going to repeat this again and again, until the nest starts to take shape,' Iain tells me. The stickleback runs through the routine a few more times; then the video cuts forward an hour or so, by which time the nest is beginning to look like a tiny archway on the bottom of the pond. By now the poor male is looking even more flustered than he was at the start. A great deal more fraught. His fins are a blur now. He starts to pump them vigorously, wafting water against his construction as if to test and test again every weak-spot. He tries it from every angle, hovering, zipping from one end to the other. And then . . . there is a momentary contentment, perhaps.

The screen goes blank once more. This scene is over. Another caption appears on the screen: *Male courting a female.* Again, the familiar tank appears. The male's nest is now finished (it looks a little like an igloo, I suppose), and somebody has popped a female into the tank. The male looks more edgy than ever before. Wow, he's positively buzzing. He swims in sparky zig-zagging jerks from one edge of the screen to the other, before hovering silently about an inch from the gravel, both eyes facing firmly in the female's direction (which is difficult, if not impossible, for most fish). She hovers near the surface, minding her own business. It being spring-time in the video, her belly has become harp-shaped, her eggs ready for the suitor below, which, astoundingly, she appears not to have noticed. The male looks furious about this, and zips and sparks toward her and away, to the nest and along the perimeters of the tank. His mouth pumps water in and out rapidly, and his tail fin waggles as fast as the wings of a bumblebee. Still she doesn't look. Angry at her lack of attention he drives himself toward her, hovering over her body and nudging her about a bit, left and right.

'Watch carefully . . .' Iain suddenly becomes more animated. 'The male gently pricks the female with his spines, under her chin,' he says with a hint of amazement. Blink and you'll miss it, but he's right: the male stickleback does indeed appear to spike her with his spines. As if to prove it, the video shows a replay of the behaviour in slow motion. Yep, he definitely jabbed her. TWACK! She flinches and immediately takes a momentary interest in this surprise interruption.

'If she likes the male, at this point she might follow him to the nest,' the video commentary intones. Even now, as it looks as if the female is cluing up to what's going on, the male is playing it just too damn eager. He thrashes into her wildly, and edgily thrusts himself this way and that, beckoning one second, stabbing the next. After a minute or two of this, she's at his nest and ready to view his wares.

Following what seems like a lengthy period of consideration, the female pops herself down and into the nest's arch of threads, pushing her big swollen body in to have a good hard look. But then, she pauses . . . she's stuck. The male squeezes his head into his nest next to her and starts to vibrate. 'Now the female is inside the nest,' the voice-over explains, 'the male begins to tap vigorously on the female on the sides, to stimulate her into egg-laying.' He pecks, throbs and trembles all over her behind. She lies there motionless, hidden by the lattice of long black threads, which he obsessed about for so many days. Then – POW – she wiggles her way up and out of the nest, and she's gone. Without wasting a second, he dives into her slipstream, into the nest, and waggles himself over her spilled clutch of eggs, pumping out his milt (as fish semen is known) all over the spoils. Boom – she's gone, and he's bagged his first, hopefully, of many. He immediately heads over to the female and nips at her tail, should she start getting ideas about eating the (now fertilised) eggs she left behind. The narrator of the video wraps up: 'The male chases the female away, and his long period of parental care begins.'

It's a funny one, this. Male sticklebacks are often marketed as virtuous – helping their offspring, protecting and raising them, while the females sit back and relax. I learn that last year the BBC did a special feature on them for Father's Day (in Iain's very lab). And I guess I can see why popular opinion has always leaned this way. Those females, they look . . . what is that? Stupid? Well, a tiny bit, perhaps. They look a tiny bit stupid. Stoned. Altogether not there. And the males? Well, they seem *totally* the opposite. If a male stickleback has a brain capable of assessing the physical properties of each and every piece of nest material, weighing up the right dance-moves at the right time, calculating the right time to get the red-rage, *and* successfully raising a clutch of offspring . . . well, what's up with the females? What are they packing in *their* brains?

Jokingly, I mention the word 'stoned' to Iain and he laughs. He relishes the opportunity to set me right about the less celebrated female stickleback. The story is often told from a male perspective, he agrees. 'Red elicits a response from males, so there's no question about that,' he says. 'But red also elicits a reaction from the females.'

I did not know this. Iain continues. 'Yep. In 1990, it was demonstrated that females have a very strong preference for males that are redder in coloration.' It was a nice experiment, actually. Give female sticklebacks a choice between two males, one lit by white light, one by green (which absorbs red light), and the result? Females like the red males under white light, and aren't picky under green. 'There was something intrinsically important about red that was interesting to the females,' Iain tells me. 'But the other thing they showed, very clearly, is what red might mean to a female. They demonstrated in the paper that redness is a very good predictor of parasite load.'

In other words, redness might exist, at least in part, because it allows females to avoid parasite infections. It's an honest signal, like the peacock's train in the previous chapter. Red isn't just about rage. That's only half of the story. Red is also a way for females to gauge the healthiest dads to have sex with. It's an ornament, a trait for showing off. It is sexual selection at work.

Iain gives me a crash course in how fish make red. Redness is synthesised from carotenoids, organic pigments that can be assimilated only through the food that a stickleback can find. The best sticklebacks get the most food, so theoretically they can assimilate the most carotenoids. But there's more. Male sticklebacks can 'choose' what to do with their carotenoids. They can either use them in maintaining their immune systems (the healthy option) or they can invest them in that famous red body-bling (the sexy option). Sticklebacks, according to Iain, have a choice: be healthy or look healthy. Invest too much in looking sexy and you die of disease. Invest too much in immunity and you live, but have no sex because no one notices you. Of course, the best males manage both, and that's the key to being a high-quality male stickleback. Redness acts as an honest signal to females, an outward sign of their overall superbness, and, to boot, it serves as a sign to other males to back the hell away. And these traits can be passed to offspring, too. 'From our work here,' Iain explains, 'we've shown that resistance to experimental parasite infection is much higher among those individuals that were sired by bright red males.' Good-quality males make good-quality offspring.

Female sticklebacks are choosy; they're 'aware' of the other players. Anything but mindless zombies, in fact. Anything but stoned. Iain agrees. 'They're all watching each other,' he says, squinting his eyes suddenly. 'They're all trying to work out what's going on, what everyone else is up to.' Like little Russian chess wizards, their strategic plays are calculated based on the movements and circumstances of those around them, and their own reproductive condition. Yet their brains are not much bigger than a grape seed. I find this rather incredible. Iain shares my astonishment at the fact that such little brains can achieve so much. There's a slight chuckle in his voice, as he encourages me to imagine a stickleback conundrum: 'One of the most interesting things is this. Say you're a male and you've got 100 eggs in your nest, and your nest can hold 600 or 700. Do you keep those 100 eggs and devote the next 10 days to trying to rear them?' I nod. 'Or . . .' he takes a breath, 'do you *eat* those eggs, giving yourself a big boost of energy so you can go nice and red, and display to three or four other females and maybe acquire 300 or 400 eggs?' What? Males *eating* eggs? I start to find this all a little ridiculous, but according to Iain it's true. Males can use the energy contained within eggs to boost their redness, thus potentially appealing to other females, and hopefully more than one. Again, it's all about strategy: what works best for a given male or female, at a given place and time.

And it doesn't end there. One fascinating thing about the evolutionary tennis match between the sexes is its sly backhand. After this adaptive male breakthrough that eggs can occasionally be eaten for reproductive value, evolution has now encumbered females with a wariness of empty nests. To them, an empty nest is a sign of a male with questionable sexual intentions. Females avoid them. So what's a male to do? Simple: steal eggs from another male. And that's exactly what they do: they steal one another's eggs to make their nests look bigger and better, and to make it at least *look* as if they are not egg-munchers. Forehand smash from the males; the ball is in your court, ladies. And as with most of life on Earth, the sexual game is at deuce – forever. Each evolutionary advantage is smacked back across the net for the other sex to adapt to. Male forehand, female backhand, male backhand, female forehand. Advantage male. Advantage female. Forever.

But that is just one facet of the stickleback's tale. For every impressive male attribute (or 'good dad' behaviour) there is an equally forceful backhand, played by the females in a language (like subtle behaviour or choosiness) that is harder for human eyes to read. Thank goodness, then, for Lorenz and for people like Iain. Sticklebacks are an unpredictable soap opera, forever twisting and turning in their devious plots. Good fathers? This is the wrong label to use, for she's also a master at her own game. He's maximising offspring, she's maximising quality. There is no 'good dad' or 'good mum' – only genes, and lasting strategies for spreading them. Lorenz paved the way for knowledge like this to be gleaned from studies of stickleback-kind, and the world is a better place for it.

Before I head off, Iain hands me a four-million-year-old fossil of a stickleback, which I duly paw and study. We talk in more detail about the modern stickleback, this ancient mariner, washed up into thousands of populations in our ponds, lakes and rivers after the last glacial retreat – here, there, then, now. Fifty years ago, scientists like Lorenz used fake wooden sticklebacks as tools to understand the minds of real sticklebacks, yet here and now this humble fish has its spines in almost every biological field. Where once international stickleback conferences were about little other than extensions of Tinbergen and Lorenz's work on sex and aggression, now sticklebacks are part of an enormous range of scientific endeavours: behaviour, evolution, development, photoperiodism, ecotoxicology, farming, genetics . . . 'The fact that the last stickleback conference was held at the renowned Fred Hutchinson Cancer Research Center in Seattle tells its own story,' Iain says as we walk through seemingly endless corridors on our way out of the building.

We talk some more about the conferences and Iain's students. It seems a timely moment to ask Iain about how he got into sticklebacks. 'Did you follow the cliché of a young boy, mucking about with sticklebacks in jam-jars?' I ask. 'Yeah, of course – it's what we did all summer,' he laughs. 'You know, you go out there and try to catch the red ones.' I chuckle at the simplicity of outdoor pursuits in a childhood a decade before my own, just before Nintendo and Sega's rampaging assault on the young. And today's students? 'Students now are brighter than ever, they come here specifically to study zoology – they're bright, they're all motivated,' he says with sudden seriousness. 'But I'd say that 90 per cent of them, when we first introduce them to sticklebacks, don't know what they are.' He pauses. 'And that's sad. That's really sad.'

Iain tells me about his side-project, Sticklebacks for Schools. He speaks passionately about his keenness to see every classroom have a well-maintained tank, with its own resident nest-building male stickleback. He paints a picture I rather like: a new generation of young people, watching status updates from a feisty, sexed-up male stickleback that hovers in the corner of the tank. A new generation of Lorenzes and Tinbergens watching through the fourth wall. A stickleback sexologist in every school. The idea appeals to me greatly, for sticklebacks seem to offer up so many potential treasures to future scientists, perhaps many we can't yet imagine. Who knows what else they might teach us about life and sex? Or how hindered the lives and loves of our grandchildren will be by not knowing sticklebacks in the same way as people like Iain did when they were children?

While researching this chapter, I discovered a quote from *Love-life in Nature* by Wilhelm Bölsche. Published in 1926, this book is from that era of 'loathsome harlotry' and the forbidden sex-acts of Levick's 'hooligan cocks'. It's almost laughable in how dated it sounds.

> *The male, and only the male, is the hero in the life-epic of sticklebackdom. The female, at best, is just an episode in it. The male is representative of the whole ethics of the race, existing not only as an individual pursuing its own individual aims, but as a citizen of a higher community perpetuated as a species through the millenniums. The female is really nothing but a roving gipsy, living a free and easy life without any duties on her conscience.*

This little fish has come a long way. Male. Female. Fighting, engineering, showing off, investing, stealing, making choices. Each year, living, dying, living, dying. If all of our fresh waters become poisoned or dried up in the coming century, the stickleback will live on in labs like Iain's. The conferences will continue. And these fish, both male and female, will remain forever, to science at least, nature's sex symbols. A creature that helped start a sexual revolution, and that today sheds light onto an evolutionary tie-break we are all, forever, part of.

CHAPTER THREE

Waiting for Frog 'O'

We live at an extraordinary moment in time. Of all the moments at which you could have been born, you appeared at the exact beginning of the end for the amphibians. You have been born into a class war: one class, the Mammalia, is up against another, the Amphibia. Or put even more simply, one species, *Homo sapiens,* is up against a whole legion of amphibious artisans. And unfortunately, we're winning. At the time of writing this chapter a third of amphibians are listed by the IUCN as threatened with extinction. Most species' populations are declining. Pollution, disease, climate change, loss of habitat, invasive introductions; they're victims of the usual human assaults.

This has always made me rather sad, for amphibians are such sports. The creatures that time forgot. Lovers of land, wedded to water. It's a miracle, in some ways, that such creatures have persisted this long.

The build-up to amphibian sex is relatively easy to study. After all, you only need a pond and a torch. And a forgiving partner who won't mind you coming to bed late. And neighbours who won't call the police if you spend inordinate amounts of time standing outside at night in your pyjamas, staring at frogs gearing up for sex.

I should know. I've done this for years. Amphibians are my bag, so to speak. My career started with them, and I've been blessed by surrounding myself with kind, understanding people who won't report my frog-related actions to the authorities. I have a guilty secret, though. One I'm slightly loath to admit. You see, I've never actually *seen* the sexual act occur between male and female frogs. I've seen the union, but not the deed, so to speak. I've yet to see the female actually *extruding* her eggs and the male *pumping* out his semen all over them. I've seen plenty of build-up – hours and hours of it. We all have, right? It's just that I haven't seen the proper sex; the mixing of sperm and eggs. This might surprise you. In theory, it seems like an easy thing to see. But yet . . .

I realised this had to change. I penned it in big letters on my white-board – 'JULES MUST SEE FROG SEX' – before circling it twice. I prepared my camcorder on a tripod next to the pond, ready to record evidence of this most sacred of liaisons, eager that this, finally, would be my year. Would I be successful? Well, best laid plans, and all that . . .

*

It is mid-March as I write these words, and it has been largely wet and slightly warmer in recent weeks. By now, at about this time each year, all of the frogs around here have emerged from their winter slumber and are moving each night, like an army of animatronic extras from the film *Labyrinth*, to 'their' breeding ponds. Mine is a small, raised plot in our backyard in which sits a pre-formed plastic pond, filled with rainwater from the butt. It sits right next to the front door so I can see it in my pyjamas, handily. And today it is writhing with licentious frogs. There has already been some activity. To my frustration two blobs of frogspawn float listlessly in the corner (I missed seeing it being laid) and holding onto each blob like an enormous life-raft sit two spry male frogs. They puff out their chests (slightly blue at this time of year) and look as if they mean business. One of the blobs shivers slightly. There are more frogs moving around underneath, like sharks under a boat. Perhaps five or six, maybe more.

These are common frogs *Rana temporaria,* perhaps the most numerous and widespread of Europe's amphibians. They can be mottled, green, brown, ink-stained or sometimes even pink; they'll often be nervous and jumpy, and always they'll have that familiar black 'mask' under the eyes.

Amphibian sex lives are made complicated by their common need for water. Because their eggs lack hard shells they are drawn to it, or at least most of them are. That means that, for sex, amphibians must gather together; and often this is in ponds that might be quite small, like mine. As with the sticklebacks in the last chapter, ponds can force competition between individuals, and this makes things interesting; natural selection works quickly in a scrum.

Frog-watchers might notice the development of a few quirky sexual characteristics at this time of year. Most obvious is that females grow big. Really big. Their bodies swell with hundreds or even thousands of eggs. But changes occur in the males, too, for they are victims. Victims of competition. You see, in most ponds there are more of them than there are females. Way too many. The sex ratio is skewed their way, partly because of a quirk of their sexual behaviour. Females, once mated, head off from the pond. Males don't. They stick around on the lookout for further mating opportunities, influencing the sex ratio at each pond enormously. But the dice are actually loaded even more heavily against them, because females take longer to mature sexually after metamorphosis (the change from tadpole to froglet); therefore, in every population there are more males ready for sex than there are females. Ponds become very male places indeed.

Where male elephant seals have famously evolved toward enormity and peacocks towards the preposterous, male frogs have evolved instead toward something else. Competition has forced the evolution of a most awe-inspiring sexual adaptation. It is a power cuddle. Known as amplexus, it's essentially a cuddle from behind – one which the female appears (on the face of it) to do little about. The male grabs on, and holds on to her for dear life. And he has a tight grip. A grip so powerful not even a skilled plumber could pull them apart. It's a sight you're likely to have seen yourself, so common is this behaviour in spring. The male frog's front limbs appear to lock in place, like novelty grabbers at a funfair. Often, she can't shake him. And neither can rival males, at least not easily. The next time you see this have a closer look at the male's front feet and you will notice his 'nuptial pads' – special suckers on the first digits, to aid his grip. Sex's driving gloves, basically.

The power cuddle is evolution's gift to many male frogs and toads, wedded as they are to finite, competition-rich habitats like ponds and lakes. Winner takes all. But it's not all about cuddling. Many frogs invest in other strategies to stand out from the crowd. As a class, many amphibians have had to become a rowdy bunch to make their advances known, evolutionarily speaking. Shouty, fighty, restless, yowly: you name it, these males are doing it somewhere loudly, right now. Frogs in particular. The loudest frog of all is the common coqui, a brownish Puerto Rican frog with large dark eyes like marbles. Its two-part call (predictably 'CO!' and 'QUI!') is as loud as a lawnmower going off in your face, much to the fury of residents of Hawaii, throughout which this non-native frog is currently spreading (in Britain, we have our own noise-polluting invasives, the green frogs; they too are said to sound like lawnmowers, though personally I consider them to sound more like a mallard blowing a raspberry). The 'CO' and 'QUI' noises that the coqui makes are interesting, since they have different purposes. The 'CO!' is, apparently, aimed at other males. It encourages them to scarper. The 'QUI', however, is targeted at the females, luring them nearer. Like birds, their calls mean different things depending on who's listening: 'FEMALES COME HERE / MALES GO AWAY', in other words.

Amphibians are nothing if not diverse. There are some freshwater habitats, next to raging streams and rivers, for instance, which are too damn loud for a colloquial 'QUI!' or 'CO!' to be heard. So in places like these, some frogs have hit upon a wonderful behaviour: they simply wave at one another. A little wave from one side of the stream to a female on the other. 'Hey . . . you over there!' [Wave] 'Hey . . . Hi!' [Wave] 'YEAH . . . HI!' [Wave]. If aliens ever choose to visit Earth, I sometimes wonder whether something like this, rather than the seeming majesty of us primates, will pique their interest ('YES, CALL HOME. THERE ARE WAVING FROGS ON THIS PLANET . . . YES, THEY WAVE').

I love frogs for reasons like this. Of those that I know of, perhaps the most impressive example of frog adaptations for tackling competition is the Otton frog *Babina subaspera,* a resident of the Amami Islands off Japan. It has evolved an extra finger, a 'pseudothumb', which bears a sharp, Wolverine-like spine that can be thrust into competitors during pre-sex wrestling bouts. Stick that in your pipe, rival males, or so to speak.

But hang on. Stop a second. There's a trend here, wouldn't you agree? Males doing this. Males doing that. Males grabbing. Poking. Slamming one another off females. Waving. QUI-ing and CO-ing. What, in God's name, are the females up to? It's that familiar story once more, which we touched on with the sticklebacks in the previous chapter. Males seeming to do all of the work; females apparently without much to do. Could life really be that simple for the female frogs? Predictably, no.

It's true that female frogs often lack much of a voice, and it's also true that she may suffer the indignity of having to carry around a male in amplexus. I suppose her role sounds passive, then, but of course it isn't. Genetically, she is most likely to gain from mating with the strongest and most determined male. It pays for her to court a good hugger, in other words. Good hangers-on are strong. They have value. Having sex with strong huggers makes strong-hugging babies. In theory, such traits benefit her young sons (indeed, there's evidence that if laden with a sub-standard male, a female common toad may actively seek out and encourage competitive interactions between males, presumably to get the best, strongest male going, and thus acquire the finest genes for her young).

And, of course, she's making choices too. A host of studies have examined, and found, elements of female choice in frogs – many based on male calling. Some females like long loud calls, others prefer piercing shrill ones that fire out in short bursts. Each male has evolved to make his call the most efficient yet effective call going – at least to the eardrums of a female of the species. But she may drive the evolution of such traits by holding all the cards – namely, her prized eggs.

But there are such things as feisty fighty female frogs too, most notably among females of the midwife toad (though they're actually frogs, and don't ever forget it). Here, since the sexual conquests of male frogs are hindered by childcare (males wrap fertilised spawn around their hind legs and protect it from drying out), it is the females that have evolved the competitive streak. Often, there might be more egg-laying females than there are available males, a situation that becomes problematic. The result? As one might expect, seeing females grappling with one another for access to a male is commonplace in midwives. Indeed, in the case of the Iberian midwife toad, the females are the ones that call, distinguishing themselves from their rivals in an attempt to locate a suitor.

But it's not all loud and raucous in the world of amphibians. In newts, things turned out a bit differently. Female newts can absorb sperm in little packets called spermatophores, and this small difference in anatomical 'design' means sex plays out very differently. In newts, males don't need to grab on and hold tight, they need only to encourage the female to take aboard their spermy cargo. This changes everything. While males still compete for this honour, there's none of the physical struggle. There are no killer cuddles. Newt sex is slow and more ... well paced, somehow, involving lots of showy posturing from males to females. Males use crests, if they have them, to display their vigour, and they use their tails to fire little eddies of fragrant, pheromone-rich odours the female's way. Eventually the male deposits his spermatophore on the floor before coaxing her over it. If she accepts his proposition, she lines up her cloaca with the spermatophore and shuffles it up into her body – hey presto, job done. The male then goes off and tries his luck elsewhere; the female is free to get on with the long task of wrapping her (now fertilised) eggs among pond leaves and detritus.

This doesn't mean that it's a cake-walk for male newts. It's still a battleground for them out there, it's just that, instead of hand-to-hand combat as in those frogs, the male newts settle for ... well, pheromone-filled water pistols, I guess. And the good news is that, like frog mating, it's a behaviour that nearly anyone with access to a pond can watch. Just get a torch and shine it across the bottom on a spring night and you're likely to see them there, prancing around, sizing each other up, and firing pheromones at one another in much the same way as they probably did in the age of the dinosaurs.

Newts fit within a larger family, the salamanders. The largest of these are the giant salamanders, which come in a big Chinese version and an almost-as-big Japanese one. These surely are some of the world's most fantastic creatures. They are so enormous that only a handful of men on Earth appear to be able to hold one without grimacing like Hulk Hogan (at least, according to my Google Image search). Up to almost 2m (6ft 6in) long, they are like the world's weirdest novelty doorstops. They are evolutionary outliers, really, and on their last legs too, particularly the Chinese giant salamander, sadly now saddled with the label of Critically Endangered. So threatened are they that, as with pandas, scientists are becoming increasingly interested in their sex lives, particularly with regard to making them have sex, indoors, for the sake of repopulating the landscapes in which they once thrived.

Giant salamanders differ from most others of their kind. They breed externally, as frogs and toads do (but unlike most salamanders), the females extruding eggs that the males then fertilise. And, like many frogs, males have evolved to become sexually brutish. Where male elephant seals compete to become 'beach-masters' (reproductive powerhouses that protect harems of females), giant salamanders compete to become 'den-masters'. Instead of protecting harems of females, they aim to protect their eggs, in a specially prepared enclave in the river bank. To male giant salamanders, these burrows are an obsession. They fight and wrestle (sometimes to the death) over such prime real estate. And they strategise. Some males choose to sit quietly in their den, and chase off any other males stupid enough to poke their head in; some instead choose to prowl and patrol, investigating other, potentially better, nest-sites. Some, normally the older, burlier ones, take over multiple sites. They encourage females to check out their quarters and then, should one choose to lay her eggs there, the male will fertilise them, and protect them for the next month or so from hungry interlopers. Occasionally, a female will be halfway through laying her eggs when a handful of other males burst into the burrow, squirt out their sperm, and clear off again, much to the enormous anger of the den-master. And, like many amphibians of temperate regions, the giant salamanders are annual breeders. They time their reproduction to ensure that their offspring hatch into a land of plenty. Namely, spring.

Let's look at spring. We tend to think of frog breeding (and their fabled breeding migrations) as an early spring phenomenon. But seasonality has become a key question within sex science in recent years. After all, how does a frog really know when spring is? What is it about the coming of spring that they notice? What do frogs know about the Earth's transit around the sun? These are interesting questions, certainly, and the answer is perhaps harder to pinpoint than you might imagine; it is a question that still occupies the waking thoughts of many of herpetology's greatest minds.

Along with a host of other vertebrates, it is likely that frogs' sexual seasonality has something to do with the Earth's tilted spin. Or more correctly, the gradual impact that this tilt has on day-length each day: the so-called photoperiod. Though temperature, diet, social interactions and lactation time may influence mammals, it is the photoperiod that leads the sexual show – something that farmers (and racehorse breeders) have long understood. Get a ram. Stick it in a shed. Mess with the lights over a series of days and weeks, and you can make his scrotum inflate and wither and then inflate again like a pair of bagpipes being played in stop-motion. In that ram's cells, daily oscillations in protein abundance (influenced by the blue light of dawn and dusk) are being measured without any conscious thought; a measure of changing day-length is being calculated. The circadian clock is ticking. Eyes are the receptors of this information, but not the familiar rod or cone cells. Interestingly, it's all done through another set of cells: the photosensitive retinal ganglia, part of an ancient photoreceptor system that responds to a photopigment (melanopsin) that is sensitive to these exact bleary light conditions. This is running the show ... or at least introducing the dance (and yes, of course, these cells are in *your* eyes, too).

But it is time to move on. I'm aware that I've done little else than tell you lots about amphibian sex. Is it time to actually see some? Back to my pond, then ...

It has been a few days now, and the wet humid nights are really setting in. The pond still throngs with the night-time activities of a host of frogs, which I observe for hours at a time each night. Rich, earthy smells fill the air that surrounds the pond. Nature's morning breath. It's a smell of growth, and I notice it each year before promptly forgetting all about it for another 12 months. This is my favourite time of year.

'And how is the frog sex challenge working out?', you might ask. Well, I've had the camera set up for the last four nights – pointing at the spawn raft that sits in the corner of my little garden pond. Before pressing RECORD each night I have crouched down by the water's edge, eyeing up evidence of the previous night's high-jinks. Now four spawn-blobs sit there. One seems suspiciously small, which means it was only recently laid (spawn swells up to its normal size after a day or two). Today, there are at least five males scrapping over the seeming privilege of getting to sit atop the growing raft of eggs. To the left, the only female present is suited up with a male. He is clamped on tight, his front legs clutched vice-like around her waist. Every ounce of his strength appears to be focused on the act of holding on, like some sort of parasite (some, in fact, would say that is exactly what he is). Even his eyes are pulled into his face, drawn toward her body, somehow chipping in to his body's aching fight to keep her from another. The female has almost a rose-orange tint to her, her colour diluted by her ridiculously stretched skin, laden as she is with hundreds of eggs. What is she waiting for? This is the third night they've been paired, and still nothing. Still no movement. If it weren't for the little rhythmic pulsing of her chin I'd have assumed she was dead. When will she go for it? How will he know, lodged tight behind her body like that, when she's getting ready to squirt out, to extrude, her eggs? What, exactly, are we waiting for here? Another day passes.

Questions like these have bothered frog-watchers for years. Frog migration and the timing of their sex lives is rather mysterious; the journey from land to pond – ready for sex. For years, the jury was out about how exactly they find their way to breeding ponds each year. Only in the past few decades have scientists begun to piece together the cognitive talents involved, and they are certainly impressive. Most herpetologists seem to agree that in common frogs, at least, it might involve algae, or at least the smell of algae. This 'algae theory' was first put forward by Maxwell Savage, champion frog-scientist of the middle 20th century and a man who knew his frogs, or at least those within a two-mile radius of his study patch in then-rural Barnet. His theory explained a quirk of frog sexual behaviour – weather. Frogs are very pernickety about the weather. Usually only a succession of wet, mild nights will see females release their spawn; Savage's deduction was that they simply hang around waiting for the algal crop to ripen. I can vouch anecdotally for this. If you ever disturb a frog on migration, it certainly seems to know where it's going. If you come across one with a torch, rather than flee it will nearly always carry on in a given direction, so single-minded does it become at this time of year. They're homing in on something. And they do appear to move upwind toward a pond, too.

But there is more going on. We know that frog eyes and ears are good for orientation when near the pond, but other senses may lie behind their ability to home in on a chosen breeding site at the right time of year. Some amphibians use the sun to find their breeding pond, some use polarised light (as do bees), and some might even be able to read the moon and stars, or perhaps orientate themselves via the Earth's magnetic field. We still aren't sure which of these abilities our common frog may be using in combination with smell. Even more impressive, perhaps, is the fact that there must be an element of memory, too, for frogs sometimes turn up in gardens where ponds have been removed but in which they spawned in previous years.

Things like this really lift my heart. I love a good mystery. I find it pleasing that the frog – a common animal, a garden pet, one of the most studied lab animals in the world – still has something shadowy about its sex life; they can still keep their secrets tight within their little heaving chins. And that, I guess, is why I've had to resort to using a video camera to see them actually get down to it. This may sound perverted, no doubt, but I at least have a purpose. It's something that I've never seen. I want to see it. Frog sex. That big moment in a frog's life.

Though the build-up to frog sex is easy to observe, the act itself – female extruding her eggs while the male fertilises from behind – is harder to see, as I outlined earlier. Frogs tend to do it in the depths of the night, and certainly not while being watched by odd (if honest) primates like me. In 15 years of frog-watching, it remains an experience that eludes me. My video camera has been running each night for a week now. Perhaps last night I finally got it on film? There was a new blob this morning, an observation that bodes well.

I walk inside, open my laptop and fast-forward through last night's video. On the screen the frogs skate and skitter in jerky movements from one side of the pond to the other. The surface of the water seems to wax and wane with them, like a time-lapse of the tides. Sometimes there appear to be eight or nine, sometimes just one or two. I stop the video and press PLAY. In real-time, mostly the frogs sit still with their heads resting on the surface of the water, their throat-pumps creating tiny rhythmic waves across its crystal surface. Occasional groans come from somewhere deep within the spawn-mass, like the intestinal workings of some enormous monster out of shot. The croak of the common frog surprises me each year in how different it sounds to the Hollywood croak, which echoes from gift shops and Disney stores the world over. Our frogs sound more like footsteps through thick snow. Or marbles being intensely rubbed together. A frictional croak, somehow. A creak.

The video continues to play. There, on the screen, sits one small male, one pair in amplexus and one large male straddling the three spawn blobs that float in the corner of the screen. They sit there, unmoving, as I fast-forward through the hours. The pair in amplexus bob like corks, barely moving but for the female's chin. The male on her back has his arms firmly clamped under her throat, a position that looks incredibly uncomfortable for the female. Perhaps this really will be my night?

I once worked on a frog help-line (yes, seriously), and this phenomenon of amplexus threw up lots of enquiries from the public, especially the question of 'IS HE HURTING HER?' I could understand people's concern about this, especially since females sometimes die during the males' obsessive attempts to clamp onto their backs. No one likes to see animals strangled, not least by one another, and so it was for the frog-watchers who would call me on my help-line. Mostly I was honest with enquirers. For both males and females, breeding is incredibly demanding and often both sexes will see casualties, I'd tell them. In the morning during early spring it's not uncommon to find them spent, their eyes opaque, lifeless, like washed-up salmon. As we see occasionally in nature, a female may end up paying with her life for the best genes; a male for the most offspring. If they've managed to breed beforehand, though, they remain genetic epitaphs in future generations, and that's what persists. Live or die, evolution maintains an interest if they, at least, get two things in the right order: survive, then breed. Genes that manage that keep on going.

Indeed, in some frog species, even death won't stop them breeding. In 2013, males of a tiny Amazonian frog were observed to *squeeze* eggs from dead females, which they then fertilised in the water, a kind of evolved behaviour the authors term 'functional necrophilia' (in your face, Adélie penguins!). No one's seen it done in our common frogs, but who knows? If it works, it works, I guess – for females and males.

I fast-forward more quickly now. The male frogs appear and disappear jerkily like whack-a-moles in different parts of the camera frame: three, four, one, five. Suddenly there are seven. They wrestle and writhe on the spawn mass. I find it hard to believe this all went on a matter of metres from my front door while we slept upstairs. Yet still the pair in amplexus hang there, almost motionless, in the corner of the shot. When will they have sex?

I press PLAY again and watch them both to check that the female is still alive. The counter in the corner says that it's 01.42 am. Our cat looms out of nowhere in the back of the shot, like something from *Alice in Wonderland*. He has a quick drink and retreats back into the dark. I whizz through the video a bit further, hopeful that this will be the night. Their night. My night. But, no. The screen goes blank. Battery expired. The end. Alas, it is not to be. The counter in the corner freezes at 02.34 am.

They ended up having sex. As I mentioned I was greeted by their fresh blob of spawn the following morning. Sods. They outsmarted me again, outwitting my attempts to film their sex. Foiled by bad batteries, and ill-adapted mammalian technology. A rare win for amphibian-kind.

A few days later I revisit the pond at the day's end. It's about 10°C, and there is a muggy mist over the pond. In there now, the blobs of frogspawn jiggle ever so slightly in the humid breeze. The eggs are growing. Each a dividing mass of cells which will, within days, ripple and stir as new life. But the old life has gone. The adult frogs are nowhere to be seen, for this year at least. The next time they return will be when the Earth is back in this familiar part of the Solar System, the part where my hemisphere reaps the benefits of a photon deluge, and the ponds become alive with the promise of blessings – provided care of a single star and a planet with a wonky spin.

Of course, I'm sad that I didn't get to see the sex. I'm upset that these backdoor animals dodged my sexual interrogation again. Gutted that an animal, among life's most ancient and once prosperous, could get one over on me (A MAMMAL!) once more. But, predictably, I love them for it, too. I can't help myself. It's fitting somehow that my inability to observe their sex should continue in a year when animal sex is pretty much all I've been thinking about.

If there's one thing I've taken on board from this whole experience it's this: for all we talk about sex, the act itself can be difficult to spot – hidden in a compost heap, in a nest-box or in a pond like mine, close to my front door. The actual mixing of sperm and eggs – the business end of sex – remains an often secret encounter, hidden, kept away from prying eyes. And all of it timed to perfection.

These frogs. They see spring coming. They detect its imminence. They predict it. They bank on it. Their bodies are sensitive vehicles taking in seasonal cues, revving up engines for breeding – smelling out that algae, gauging when there might be just enough to support their offspring. Ready for competing, shouting and, if need be, waving. And cuddling. Each year, in my pond, those frogs are quickest off the mark. They are a sign of the oncoming spring. A measure of Earth's transit around the sun. I raise a little toast to Copernicus, and to the rest of life's astronomers, before heading to bed with nature's morning breath ringing in my nose. The season of sex is here.

CHAPTER FOUR

The Cloaca Monologues

Type 'Which animal has the biggest penis?' into Google, and you'll be able to take your pick from the 8,500 search results that come up (well done, yes – it's a blue whale). Now type in something different. I don't know . . . say, 'which animal has the biggest vagina?' Fair question, you might think, but wait: just 67 results come up, at the time of writing this book, at least. 67? Isn't that a little strange? Why the discrepancy? It seems staggering.

To get my head around that question I decide to pay a visit to Chatsworth House in Derbyshire. Not the first place to think about vaginas, you might think, but worth a visit nonetheless, for reasons I'll explain later. Famous for its luscious riverside vistas, its historic splendour and the masses of tourists eager to see the spot where the BBC's adaptation of *Pride and Prejudice* was filmed, Chatsworth is the perfect place to watch the spring come in. And the perfect place, too, to see the complex seasonal behaviours of one of its most numerous and charismatic residents.

I arrive early to make a day of it. By 10 am the car park is already half full. Balancing an absurdly wide Americano, I pull up a chair in the stable quarters (now a café) and stare at the rather regal raised pond in the middle of the courtyard. Tourists bustle every way I turn. I'm not like them. I'm here for a different reason. A fountain sprays water like a geyser, up and into the rousing spring sunshine. I have a double layer of hoodies and a scarf (the spring is bright, but I can still see my breath), plus my notebook. I am here to take some field notes about a creature that all sex-writers must make an early-career pilgrimage to see. There are plenty of people here, but it isn't their genitals that I'm immediately interested in. No. Me? Please, officer, I'm only here for the ducks . . .

Chatsworth is a lovely place for mallards. Many hang around the river, or lollop around on the long, well-kept lawns. If you visit in early spring the place is home to hordes of them, males preening their greener-than-green heads, and females performing a complex chemical process that turns scraps of bread into eggs being prepared deep within their bodies (courtesy of medullary bones just like those found in 'B. rex', if you're asking).

It's now later in April, and spring is coming to a stuttering start. Sitting in that breezy courtyard I scribble a few bits and bobs in my notebook. My binoculars are on the table, ready for any mallard that might sweep across the Tudor skyline. There is something lovely about Chatsworth. People who visit here appear happy by default, confusingly cheery – much like those who have National Trust stickers in their car windows. They wear visors. They amble. They are happy simply to be alive. They are the sort of people who carry breadcrumbs in their handbags just in case there should be an opportunity during the day to feed wildfowl. And the ducks here know it. On many of the occasions that I have visited in the past, there have been hundreds of them.

But not today. The place seems entirely devoid of ducks. Patiently, I wait for a few minutes. Then my first one appears. She's padding around twenty metres away or so. From across the courtyard our eyes meet. Hers are beady. Seeing me as a potential dropper of crumbs, she waddles over. With a slight limp, she weaves through the legs of the numerous plastic tables that fill the courtyard. She's locked on. She's homing in. Her confidence seems to put the little dogs on leads in the courtyard slightly on edge. When she reaches me she stands stock-still centimetres from my legs. She lifts her head and looks at me expectantly. I pat my pockets as if she's asking for spare change that I don't have. 'Sorry,' I offer sadly.

This has worked out rather well, though. For her closeness allows me to take her thoroughly in. She is beautiful. Up close, one can see how intricate her colours are, a patchwork of browns – tan, hazel, mahogany, mousy, sandy, tawny – arranged like highlights across her body. The colours expand and squeeze together as she breathes, as if I was looking at stripy wallpaper while drunk. She momentarily scans the floor under my table for crumbs and flicks her head back up at me, tilting it slightly; that classic glance – the vertebrate question mark. Her feet point comically inward at one another, as if she's suddenly come over all shy. And she has, I suppose. That camouflaged coat screams out that she often wants to be left alone. It tells a story of protection of eggs and young, of blending into the background. Avoiding attention. That's what she and her kind do best: avoid attention. Whether it's from foxes, peregrines or, worse, those feisty male mallards, the drakes. For they are integral to the story of her vagina, in a way that foxes and peregrines certainly are not.

She lets out a polite *quack*, and moves off to another table.
I look around. 'Why doesn't she have a suitor?' I think. I
wonder about this for a while, because I know a bit about
ducks, and I know that it's rare to see female mallards on
their own at this time of year. I scan the floors beneath the
tables of other punters and spot nothing. No male anywhere.
Is he by the bins? No. On the roof? No. Hanging around
the gift shop? Nothing. Could she really be on her own?
Then – THERE! – in the middle, standing on the edge of
the fountain, is the drake, watching her. His head moves
slowly and mechanically, following her movements like a
CCTV camera. People and their dogs stroll right past him,
but he keeps his glare firmly locked onto her. He looks
splendid. Drakes, all of them, look magnificent in spring;
they're one of those birds that, if you saw them in any other
country, you'd pause and admire or take photos to show
your friends and family back home. The green sheen across
this one's head and neck is so shiny and metallic you can
almost see the reflections of passers-by in it. And God alone
knows what biological hair product gives it that kiss-curl
tail. I stare at him through my binoculars. His yellow beak
clacks a tiny bit. I can't hear him, but it looks like he's letting
out a suspicious noise in the female's direction.

The things that ducks hide within their cloacas have
garnered remarkable interest from scientists in recent years,
and I've enjoyed throwing myself into their research once
more over the last few weeks. Generally speaking, according
to the literature, it seems that her vagina hates him.

Now, wait. I feel I need to get something straight before
we progress any further, for I know that this is an area within
which vagina- and penis-pedants lurk. For there is a good
argument that vaginas and penises are mammalian things,
and that using such words with regard to other vertebrates
does a disservice to the evolutionary mechanics and history
through which genitalia have come to be arranged. Is the
female duck's genital pathway really a vagina? Well, yes and

no. Technically it's a pink fleshy tube (the lower oviduct) that sits within her cloaca, through which sperm can go in and later eggs can come out. Same . . . but different. Anyway, for ease of communication in this book, I'm comfortable with female ducks having vaginas and male ducks having penises; pedants be damned.

And so, let us now go back to that duck's vagina. It is a majestic and paradoxical organ, for it appears to hate all of male duck-kind. It's like the female has a default penis-rejecting mode in her vagina that only she can override. Long and spiral-shaped with a series of cul-de-sacs, her reproductive tract is built like a booby-trapped Inca temple. And for the same reason, no less: to stop interlopers grabbing at treasures they have no right to grab.

You may be aware of the lurid behaviour of some male ducks, because they don't exactly tone it down for human eyes. The basics are this. Pairing of mallards normally occurs in October, November or December, and the two birds hang around together through the winter months. When spring shows up, the female lays her eggs and the male disappears. Visit any big pond in spring and you'll probably see him; his 'job' done (for he does not assist with rearing offspring), he's free to do as he likes. And therein lies the problem. With all these horny males around, any remaining unpaired females (including those that may have suffered a loss of their first batch) swiftly become a scarce sexual resource to the males. Competition becomes high. And competition is, as I mentioned with those frogs in the previous chapter, a powerful shaping tool for the evolutionarily absurd. These male ducks are desperate, pumped up on hormones, and, worse, they're everywhere. There's too much competition. Things get crazy. Most duck-watchers will be used to 'attempted rape flights' and 'rape-intent flights' at this time of year. Not pretty for human sensibilities, that much is true.

But, hold on . . . 'rape'? Before we go any further I have

something I'd like to say. I have a bit of problem with using the word 'rape' in this context. Though, as you have noticed, I am loose with my animal descriptives and flirt merrily with anthropomorphic language, I find use of the word 'rape' uncomfortable when talking about animals. Though it is used occasionally in scientific literature, to me 'rape' is not a scientific word – it's something different, a word that deserves to have all the negative connotations it infers intact. Using 'rape' when describing animal behaviour threatens this. It dilutes it. To counter this, some scientists instead prefer 'forced copulation', and I'm one of those. So there. Now that that's out of the way, please allow me to continue.

So, where were we? Ah yes: male ducks. Forced copulation: check. Homosexual copulation: check. Inter-species forced copulation: check. And, like the Adélie penguins, they will mate occasionally with the dead, too. The paper in which this nugget of information was published is handily titled 'The first case of homosexual necrophilia in the mallard *Anas platyrhynchos*'. The abstract, which is refreshingly succinct, reports that:

*On 5 June 1995 an adult mallard (*Anas platyrhynchos*) collided with the glass facadde [sic] of the Natuurmuseum Rotterdam and died. Another drake mallard raped the corpse almost continuously for 75 minutes. Then the author disturbed the scene and secured the dead duck. Dissection showed that the rape-victim indeed was of male sex. It is concluded that the mallards were engaged in an 'Attempted Rape Flight' that resulted in the first described case of homosexual necrophilia in the mallard.*

All of this intense competition has, on the whole, done strange things to duck genitalia over the generations. Allow me to start with a description of the male's equipment.

There are some things in life that, once you've seen them, you can never un-see. Exploding duck penises are one such vision. I have seen my fair share of slow-motion exploding duck penises recently, thanks in part to the hard work of Professor Patricia Brennan of the University of Massachusetts Amherst. The videos she has produced as part of her research on Muscovy ducks have almost reached viral status on YouTube. They show, in super-slow-motion and close-up, male duck penises ('pseudopenis' is perhaps a better word) expanding to full size and ejaculating. These videos really do exist. And they're bizarre, and deeply fascinating, to say the least. In them, slightly shaky human hands hold a duck from which sprouts a strange, fluid-filled tube that looks a little like an inside-out water balloon filling with water (which is *kind of* what it is – ducks fill their penises with lymph fluid). It continues to lengthen hypnotically until suddenly it bends and twists into an unruly corkscrew. Then . . . TWANG! The tube suddenly jolts as the penile skin reaches its maximum. Then out pops a little semen from the tip. Done. Another shaky hand comes slow motion out of nowhere and holds a ruler up to measure the whole jangling appendage. It's just over six inches long. And then it fades to black. It's a video you have to watch two or three times to really appreciate. It's like a slow-motion balloon-animal show. I can still see it in my mind's eye even now. I can't shake it. The slow-motion quality of the video makes it move in the same way an Etch-a-Sketch drawing might take shape.

In real life, things move much more quickly, of course – the whole process of a duck 'erection' takes less than a third of a second. It flings out of his body at 1.6m (5ft) per second, only a little slower than a party-popper going off (BANG! AND THE EGGS ARE FERTILISED!). This is a penis that wants to conquer. And it is the drake's weapon – an emergent quirk of living a lifestyle where competition between males often becomes incredibly intense. If male ducks could speak, you can almost imagine them saying, on ejaculation, 'I WIN!' But in evolution, of course, too much winning can start wars, as we saw with the sticklebacks earlier. And that's what has happened, and where we must take this story now. For the females have fought back. With style.

A female duck's reproductive tract is anything but a nice little hole for an exploding pseudopenis to fit into. It is corkscrew-shaped, much like the male's anatomy, but here's the thing: it spirals in the *other* direction, making it almost absurdly non-compatible with exploding duck penises. Not only this, it also has out-pockets and dead-ends. It really *is* like an Inca temple. It is ridiculous. A masterpiece. Evolutionary art. And why? It seems that the females have evolved this complex genitalia to block unwanted male advances. In other words, they have evolved a mechanism that allows them a degree of choice in who fertilises their eggs. According to research, one third of all duck matings are forced by the males, yet only 3 per cent of eggs that females produce are fathered by such forceful drakes. The females have evolved the controlling hand, in other words.

Patricia Brennan owns this story. She was first to shed light on exactly how effective the female tract is at battling unwanted duck penises. And how she did it is brilliant. She made glass replicas of female duck anatomy and watched, in slow motion, what happened to the male's explosive penis as it went through them. As Brennan had predicted, the counter-clockwise spiral of the female reproductive tract slowed the expansion of the penis. When a male duck ejaculates, its sperm inevitably remains only in the lower reaches. It can't get into the deeper parts of her labyrinthine loins, where the precious eggs are. He's knocking on a closed door, so to speak. But she can loosen the 'walls' of the spiral when she copulates with a male she prefers, allowing her desired suitor's penis in a little further, past the zone of containment.

It's not that she doesn't *want* to get fertilised (such anatomy would quickly disappear from the gene pool if this was the case): it's that she has evolved an anti-spiral because it allows her the choice of the fittest individuals. By loosening her reproductive tract, she can allow the best males in. Her anatomy allows her to choose the highest quality males, thus increasing her reproductive fitness and explaining how genes that create such vaginal vigilantes could spread in future generations. Her vagina is adaptive: it pays out by making the best possible offspring, which go on to raise their own best batches.

Evolutionary biologists are used to the notion of the Red Queen, who says in Alice's *Through the Looking Glass* that 'it takes all the running you can do, to keep in the same place'. Duck genitalia are a good example of that principle. A sexual arms race between males and females, essentially. He wants the most offspring, so he evolves an explosive penis; she wants the best offspring, so she evolves a vagina that rebuffs unwanted advances, and allows her, instead, a degree of choice, based on other factors.

I sit and ruminate on this for a while in the Chatsworth courtyard. It's a wonderful story that has largely been hidden to us until quite recently. A sudden cold wind shakes me out of my daydream. The female mallard is now on her second lap of the courtyard. She returns briefly to my chair, tilts her head and gives me that same questioning look. I cave in and give her a bit of my cookie, which she dutifully nibbles up. The drake watches me from the other side, eyeing me up suspiciously. I feel momentarily as if I've been caught cheating with her. Time to grab up my things. Nice as it is to have a coffee with a duck, this is all a little tame. I had imagined this place to be a hive of scum and villainous sex acts between warring male and female mallards. But no. So far, my encounters with these ducks has been far too disciplined. I decide instead to wander the grounds to see if I can spot any more sexual liaisons between ducks. You know, as you do . . .

I walk down the hill, past the car park and through the endless ribbon of incoming cars. For so long home to lots of wintering mallards, the place currently seems to house very few. None, actually. Or none that I can see. It being spring, most are probably rearing young in more secluded spots elsewhere. It's largely silent. I carry on walking down to the river, and then ... overhead, I suddenly notice a trio of sweeping shadows move across the grass. Three ducks buzz past, their wings working like clockwork. They quack intensely. They badger and hustle one another as they fly, like relay-race runners fumbling a baton pass. One is very angry, one is very touchy-feely, and one is flapping as if its life depended on it. It is a sex-crazed lone male pestering a female, with her male guard at her side. The guard harasses the intruder from all sides, swinging left and right. They loop, squabble and spiral through the trees down to the river, where they escape my view. I hear the splash as they land, and feverish quacks ring up toward the house.

There, on that river, was where, surely, I was to see the intricacies of duck sex. Still smarting from my failure to observe frog fornication, I was determined that this – *this!* – would be my day. I tried a number of methods as I worked my way up and down the river's snaking banks that morning. I sat on the tall banks and spied the ducks from afar. I hid among the bushes, binoculars at the ready as they occasionally belted in threes and fours across the clear spring sky. In the end I settled on the following technique. It goes like this. If you have ever visited Chatsworth you will know that to get to the main house you must cross a lovely stone bridge that stands in front of the beautiful long façade of the house. This is your vantage point for spotting the sexual interactions of male and female ducks. Head to the bridge. Stand there long enough, wait and listen for the tell-tale rifle-like *quack-quack-quack-quack* coming from on high. Spin your head upwards and count the ducks – if there are three or more and they are hacking the hell out of each other while quacking wildly, then keep watching: sex is at play. At least one is likely to be a female. At least one is likely to be a male, acting the chaperone, and at least one is likely to be an archetypal desperate duck.

So, the air is one place where duck sex plays out. The other is on the surface of the water. On the bridge you can view a long section of river, left and right; an excellent viewpoint for watching action from all sides. Watch carefully. As well as the aerial encounters, you'll see wary pairs of male and female mallards ambling up and down, dabbling at the water's edge. Keeping themselves out of trouble. Paired. Guarded. But they rarely stay out of trouble for long. Keep watching and every now and then you'll notice groups of swimming males, moving like warships in formation up and down the river. Occasionally you'll see what happens when a pair, minding their own business, come across one of those sinister male battalions. The pair will see trouble brewing; they will suddenly head elsewhere, sometimes chased, sometimes not – thicker vegetation, back to the house, skyward: with binoculars you can almost see the terror in their eyes.

As morning turned to early afternoon, and early afternoon to late, it was inevitable that when sex finally happened I almost missed it. It was a fluke, really. I was watching a fleet of seven males gracefully swimming my way as I stood upon the bridge, when suddenly from behind I heard the familiar aerial honking diatribe of a male and female as they dodged and weaved under the haranguing of a male interloper. This aerial trio swooped down and under the bridge at exactly the same moment that the seven swimmers arrived. Suddenly these two troupes came face to face; a pair, an interloper and seven sexed-up submariners. Under the bridge. I knew at that moment that something was likely to take place, but what?

The noise of the action echoed up and across the grounds, threatening to spoil the serenity and charm of the place. Even the teenage car-park attendants looked momentarily bothered. Frantically, I ran round and under the bridge to

see what was going on. And there, as I rounded the corner
. . . what I saw was . . . strange. It wasn't sex (and I certainly
never saw an exploding penis), but it looked as if they were
drowning the female, almost taking it in turns to ride on
top of her. They were kneading her like dough with their
feet, as she struggled beneath the surface. They were ducking
her. Her male guard pecked and quacked furiously as the
other males seemingly each waited to have their go. His
efforts to stop them appeared futile. She was dying. It was an
image I have had a hard time forgetting, even now. It didn't
help that it took place under a bridge, like something out of
a Danish film noir. I'm not proud of what I did next. I've
heard it said that wildlife cameramen or women, working in
the Serengeti, find it hard to watch baby animals being
ripped apart by lions without stepping in. Yet they stand
firm, knowing that this is nature. I'd like to say that I had
similar zoological resolve with those desperate ducks. I
didn't. Instead I waved my arms 'GET OUT OF IT! GO
ON! GET OUT OF IT!' I screamed at those bastard
mallards. 'GO ON! AWAY!' Amid the ruckus I caused, the
female and male eventually managed a brief window of
respite, enough time to take to the wing and fly off. It was
over. The male battalion carried on downstream as if
nothing had happened.

Though thousands of people must have passed me that
day, I suspect only the fisherman who watched me from the
other side of the river for most of the day knew what I was
up to. I smiled and waved to him as I wandered off as the
sun fell low. He watched me, a cold look on his face, as I
headed back up to the car park. It was a great and enjoyable
day, apart from the 'bridge event'. I think the fisherman was
questioning why anyone in their right minds would spend a
Saturday attempting to watch duck sex. And in a way he was
right to. But here's the thing – it was a harmless solitary
pursuit. It wasn't as if I had used his tax-payer's money for
that research. He was British, not American, after all . . .

'Feds Spend $400,000 to Study Duck Genitals' came the headline in the *Christian Post* on March 21st, 2013. 'Government's wasteful spending includes $385,000 duck penis study' barked the *New York Post* not long after. Then came *The Fiscal Times*: 'Government blatantly wastes $30 billion this year' they declared, with a finger pointing firmly the duck's way. As well as raising eyebrows at publicly funded genitalia research, these pieces (and hundreds like them) also made links to recent grants made by the US government. 'Was duck penis study an appropriate use of taxpayer money?' asked Fox News. (87% responded 'No'). Bizarrely, for a few weeks duck genitalia became part of a politically charged media tornado in the United States. And Patricia Brennan, responsible for discovering so much about the evolutionary quirks of duck reproductive anatomy, was at its centre . . .

'In a relatively short amount of time, it became huge,' she explained to me in a phone call a few months later. 'In a matter of a couple of days, because of the internet, it got everywhere'. The story was out. The US government had supported basic research into the ins and outs of duck genitalia. The world would never be the same again.

But the lines were to get even more blurred, to Brennan's frustration. 'As things escalated, the stories were no longer concerned with factually reporting on the amount of money spent to fund our research; they instead became more and more outrageous. The journalists showed a complete lack of understanding of two main things,' she outlined to me. 'First, how science is funded; and secondly, in the distinction between basic science and applied science'. To Brennan, and scientists like her, basic research is a crucial tool in humankind's progression toward a better life. Her studies into duck genitalia have the potential to throw up scraps of knowledge that could be the start of a trail of breadcrumbs to riches beyond our imagination – medical, behavioural, developmental, financial. Without looking, we'll never know: the doors all remain shut. It's a notion that drives research in developed countries the world over, and Brennan defends it with a passion I couldn't help but admire. 'The whole point of *all* of this is that we build the base of the pyramid, and at the top of that pyramid are those things can directly impact on human health and economic well-being and everything else. It's all built on basic science, and duck genitalia are a part of that'.

But there was a third thing that concerned Patricia during the media furore. 'There was an underlying suggestion that there was something wrong with spending any money *at all* studying sex and penises. That it was somehow too risqué or deviant.' In a well-delivered riposte in *Slate* magazine (called 'Why I study duck genitalia') Brennan defended her research solidly, and highlighted that, in actual fact, there may be no better animal to study than a duck, anyway: a largely pair-bonding vertebrate that occasionally exhibits violent sexual coercion – sound familiar? Good for her; in fact, I hope more research into genitalia can be funded this way, rather than less. I suspect that many other common animals have similarly fantastic stories hidden away in their nether regions, if only we had the resources to look.

But I digress. It's time to get back to the story . . .

I mentioned at the start of this chapter about the apparent discrepancy in Google results, when comparing the number of animal penis stories with those on vaginas. A quick Google News search for 'penis' reveals 102,000 stories. For vaginas? A relatively paltry 17,900 (including, third from top, the headline '100 men talk vaginas'). A glance at Google Scholar, where 'penis' gets 431,000 results while 'vagina' gets 469,000, suggests that this isn't academic bias, so perhaps it instead represents a media skew? Are news editors obsessed with penises? Do advertisers baulk at the idea of vagina-chat?

Seriously, though, the animal penis literature does seem rather packed out, with oodles written about pretty much the same handful of stories. I've read too many stories about bed bugs and how they use their 'penis' (apparently) to stab into the female's abdominal wall, bypassing her war-zone 'vagina'. I've heard all about rats and their prehensile penises. I'm bored with banana slugs and their appetite for munching one another's penises off. Or that scorpionfly penises are so big they can be used as a weapon against spiders. Cats have a penis covered in barbs? Interesting, yes. But new? No. Barnacles have the world's longest penises in relation to their body, and use them to fertilise their neighbours. Congratulations! Some octopuses have a long penis tentacle that can detach, and swim unaided, toward the female. Fantastic! Heard this all before, though. Great and interesting stories all, and I don't mean to downplay them. It's just that for each and every one there is an equally brilliant story that the vagina (or equivalent organ) is telling us, if only we looked, or found better ways to investigate. The world needs more vagina stories.

A few days after my trip to Chatsworth, I learned that BBC *Springwatch* had also been visiting that day, and, amazingly, they had also been covering duck sex. *Springwatch* is the BBC's flagship wildlife programme. For several weeks each year, nature lovers revel in following the story of 20 or so baby birds, live via remote cameras, ravaged by either intense heat, floods or intense cold. The programme is a national treasure. And it seems that I had chosen, by total coincidence, exactly the same place to learn about duck sex in exactly the same week as *Springwatch*. Soon after learning this, I had a bit of correspondence with one of the researchers on the show. 'We chose it because they were studied here a few years ago,' she told me. I pretended I had also known about this study, but in actual fact I hadn't. I simply had a good eye for where ducks have sex, I suppose.

The study that *Springwatch* was referring to is important, though, because it looked at what females may rate in a male, and which males the female 'allowed' to fertilise her eggs. And this led to further studies that hint at what drove those females to take back the power, vaginally speaking. The answer is rather simple. STDs. Sexually transmitted diseases. That's likely to be at least one of the reasons those female mallards evolved such picky genitalia. The theory makes total sense: the more probe-some a male duck is with his penis, the more parasites he is likely to pick up, ergo the more parasitised is his pecker. How does evolution react? It favours females with the equipment to select for indicators of a good healthy immune system, and encourages the evolution of vaginas that are able to close off forced intrusion, thereby limiting infection and helping their own genes persist.

And what is the signal of a healthy, disease-free drake? Back in Chapter 2, we saw that female sticklebacks rate red. Well, for female mallards, the colour of choice appears to be yellow. Yellow beaks, that is. Look around in spring and you'll see them, those drakes. Green heads, kiss-curl at the tail, and a prominent yellow beak. Look more closely and you'll see that the beaks will show a host of yellow shades within a population; some are bright as a button, some are faded, drab and, well, lacking. The strategy for a female is simple: bet on yellow. It's a wonderful story; an evolutionary genital-dance that has become almost interpretative in its nature.

After visiting Chatsworth I began to notice ducks all over the place. Like the females, I judged those males; I stared at their beaks, that apparently honest signal of quality. I smiled at the females – the survivors, the ones that took back control and thrived, evolutionarily. I admired the yellowness of each and every beak, and managed almost to shake that image of a slow-motion exploding penis from my mind's eye. And, perhaps most surprisingly, I took to carrying a little bag of breadcrumbs in my man-bag (honestly).

Ducks are often mocked, characterised as comedic somehow, but we have much to take from them. They remind us that sex is a two-way thing. That every penis and every vagina has its own story to tell, each wedded to the other, with both equal in their grandeur. Clockwise and counter-clockwise corkscrews. Endless races, never won. Two players, a single story. Perhaps news editors would do well to take note.

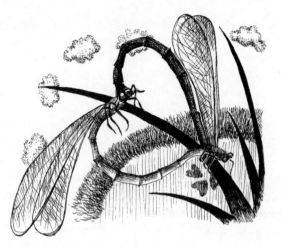

CHAPTER FIVE

The Aedeagus Complex

I have a soft spot for dragonflies. They are insects that move with real purpose, animals that know where they're going and what they're doing – a group that reached clinical excellence millions of years before any of us mammals; an insect that set upon a body plan that has had no real reason to change since before the age of the dinosaurs. And, of course, they are mighty promiscuous. Animals that fly right in the face of Darwin's big ideas about sex and his considered view that females were often 'passive' observers of male competition. For dragonflies show us a world where males *and* females have much to gain by multiple mating.

It's another of those well-told penis stories. His is rather a quirky little appendage. Like many invertebrate penises, it's shaped a little like a toilet brush, having evolved to clean out the female's reproductive tract of any other male's sperm before his own is deposited. To look at dragonflies having sex, you'd never know that this was what he was up to, though. He latches himself onto her body, pulls his abdomen downward to plug himself into her genital opening, and then gets to work. In some species the two insects almost form a heart-shape. They can stay together like this for hours. The female may deposit some eggs into the pond, but sometimes she won't. If she chooses not to, the male may eventually lose interest and zip off to try his luck with another female. She may mate again and again with other males, though, and each will attempt to scrape the sperm of his rivals from her genital passages. Neither sex appears passive; at least, that's the story the dragonfly's penis tells us. But that penis can tell us much, much more than just this.

Now, hold on. I know what you're thinking. Penises? I thought we were done with them? I've spent previous chapters bleating on about how they're over- publicised, and that their stories are well told. And they are (especially when it comes to dragonflies). But bear with me, because some penises have the potential to unlock more than just sexual truths, at least once you get your eye in . . .

In the previous chapter, I referred to penises and vaginas each telling two parts of the same story, with one evolving features for maximising quality and the other features for maximising quantity of offspring. However, to paraphrase George Orwell, some sexual parts can be more equal than others. This is arguably the case for invertebrates, for in these animals penises do have more value, at least to the taxonomists who spend their lives studying them. Without them, our knowledge of their diversity would be hindered, held back and hapless.

After my fleeting disappointment at not seeing much under that Chatsworth bridge, I decide that I need to see some sort of genitalia for my research; that I need to get 'up close and personal' with some anatomy, with the help of a microscope or otherwise. But where, and with what, should

I focus? Water beetles? Daddy long-legs? Mosquitos? Thrips? Scorpions? Water scorpions? Pseudoscorpions? Such decisions are momentarily awe-inspiring yet dizzyingly incomprehensible when it comes to invertebrates; there are just too many of them. Too much diversity. Like a kid in a sweetshop – a sweetshop where the sweets have genitals – I struggle to make my choice. Which shall I choose to focus on? Who shall I turn to for advice? I swirl in a dizzying sea of arthropods before remembering something. I remember that I have seen insect genitals almost every single day for the past three months, courtesy of an expert I follow on Twitter, Sharon Flint. Caddisfly ones, stonefly ones, beetle ones. My timeline has become almost pornographic. 'Female genitalia of *Ecclisopteryx dalecarlica*, a very nice-looking adult caddis,' says one post (her uploaded photo shows something that looks disturbingly like a snake's head. It isn't a snake's head). 'Male genitalia of *Leuctra geniculata*. Lots mating on the wooden bridge over our local river yesterday,' says the next (the genitalia look a little like a magpie feather). The day after that there's another: 'Very distinctive genitalia of female *Allogamus auricollis*', along with another photo. This one looks round, bulbous and slightly inflated, like a fat baby's head. It's an attractive genital appendage, but looking at them, they're all attractive in their own way. I have Sharon (and Twitter) to thank for making my coffee time so increasingly genito-centric. She is surely the perfect guide to talk to me about insect genitalia, and the perfect person perhaps to show me my first. I take the plunge, pick up the phone and dial her number.

*

There was a time, when I first began on this animal sex tour, when the thought of visiting a couple I've never met solely to sit in their living room and talk about insect genitalia would have filled me with nervous concern. But here I am, sitting opposite two strangers on a spotless leather sofa in their living room, drinking coffee from their finest cups, while they tell me all about insect penises. Now, five chapters in, it feels fine.

Sharon turns out to be everything I had imagined. Forthright and passionate about invertebrates, mixed with a bit of fun and a healthy touch of humour. In life, and away from Twitter, there is a hint of gothic princess about her – someone who likes to stand out, while blending in among her own. Peter, her husband, is an entomologist and former lecturer at Lancaster University. In many ways he is the Yin to Sharon's Yang – considered, reasoned and softly spoken at all times (he reminds me a little of how Theodore may have seemed in real life to the young Gerald Durrell in *My Family and Other Animals*). Together, they make a class entomological act, and immediately I realise I am in the presence of world-challenging levels of expertise.

They are much neater than any other naturalist folk I have ever met – something I admire greatly. As we talk a drop of coffee falls on the table and I pick my moment to mop it up with my sleeve, hoping they won't notice. The walls are filled with books and historic journals, the likes of which some people pay thousands to interior designers to try to imitate. On the antique coffee table between us is an old book, open ready for me to study, containing a host of diagrams of an unnamed insect's genitalia. Polished spoons, napkins and (what I would consider) the 'best' plates are out, along with a selection of (what I would consider) the 'finest' of biscuits. Now I look again, the biscuits are immaculately arranged. If ever there is a sign that I'm in the presence of taxonomists, it is this: neatly arranged biscuits.

Sharon is particularly interested in caddisflies. They've become her 'thing'. I should state at this point that my knowledge of caddisflies is pretty woeful. Like many naturalists, I'm quick to identify their larvae – the strange, maggot-like creatures that make little cases for themselves out of pond detritus – but I struggle when it comes to the adults. Being truthful, my only dealings with adult caddisflies are when I have ended up pulling their corpses out of my hair, after accidentally walking into spiders' webs littered with the husks of their dead bodies.

The adults are attractive insects, though. Often they have splendid flowing antennae and clear-veined wings kept tightly and neatly against the body while at rest. They are also blessed by a certain 'chunkiness' that I'm rather fond of. In Britain there are apparently 198 species, but most people (like me) would be hard pressed to identify more than a handful.

'Entomologists use the genitalia in the identification process, because they tend to be species-specific,' says Peter, stirring his coffee. 'For a male and female to get together, they have to be of the same species, rather like bits of a jigsaw . . .' I nod my head. Sharon continues his sentence. 'We often use male genitalia because it's harder to see what's going on in the female, because the tissues tend to be softer. Males tend to have hard parts, which are then easy to preserve and examine.' Peter chuckles slightly at Sharon's use of the word 'easy' – they both know full well that this is a topic that takes hawk-like observation skills, a steady hand and an encyclopaedic knowledge of how one species' genitalia compares to another. Insect genitalia seem anything *but* easy to get to grips with.

There does seem to be variety, though, I'll give them that. Sharon's pictures of caddisfly genitals uploaded to Twitter exhibit all of the scrawling hallucinogenia of a Ralph Steadman illustration. Penises with gorilla eyebrows. Penises with jagged edges. Hooks. Spines. Penises that look like novelty grabbers. Penises that don't look like novelty grabbers. Penises that *are* novelty grabbers. My mind spins as she shows me more pictures.

'Mind if I have a look at this?' I ask as Sharon gets up to put the kettle on for a second cup. They nod and I pop my coffee onto a coaster and lean over the book, taking it all in.

'That's a diagram of a milkweed bug's genitalia, male and female locked together,' Sharon yells out from the kitchen. As I scan it with my eye I am suddenly filled with unease. I have absolutely no idea what exactly the picture is showing. I can't make anything out. It's a mess. Genitalia? Mouthparts? I'm suddenly aware that it's taking me a while to find my bearings. It takes minutes. Peter is sitting watching me. It's taking too long. I panic, aware that my scientific cover is being blown in front of my expert hosts. My cheeks flush red and I sweat a little. On the page, far from looking like genitalia, it looks like someone has drawn a tube map from an unknown and foreign city. I'm blinded by the maelstrom of pathways, labels and lines, and completely unable to find any labels with which I'm familiar. Endophalluses, phallobases, median oviducts, the aedeagus, subgenital glands . . . my eyes scan the page, nervous that my ineptness is now definitely being noted by Sharon, who returns with our refills. ANUS! Got you! I drum my index finger on the annotation proudly. Yes, I know that one. I continue scanning across the diagram. RECTUM! Yes, I know that too (and, by God, it's enormous in the picture). ERECTION FLUID RESERVOIR! I can imagine what that does at least. But . . . the 'aedeagus'?

I finally cave in and ask my hosts for help. Peter and Sharon politely inform me that the 'aedeagus' is what I might think of as the penis. Ahhh, the aedeagus, my old friend. Suddenly the picture makes a little more sense. The penis looks like a chaotic, angry snake inside a bin-liner (correctly termed, her 'genital chamber'), and I can make out where his sperm goes and from where the eggs come. I exhale for the first time in what feels like minutes. It's a rather beautiful and elegant system, actually. And each species, they tell me, has its own complex arrangement.

According to Sharon and Peter, what led to the complexity of arrangement in insect genitalia is a matter of debate, and in each species there might be a number of competing or complementary factors. Their phalluses might be influenced by competition between males (or females), or by the number of similar species occupying a niche, or the location and time at which such competitors may emerge from their larval habitat. Seasons undoubtedly play a part. The aedeagus (or the plural, aedeagi) might be influenced by genes, themselves influenced indirectly by other factors such as diet or related life-history traits; or they might be affected by the female's attempts to ensure that only the best-quality males get access to her genetic spoils. It's likely that, as with duck genitalia, the two body parts are evolving with one another, cycling in a Red Queen battle. It may be all of these things. In some species, though, it may be none.

'He has to make sure he can insert, or inject, his sperm into the correct position as required by her receptive organs,' says Peter firmly. 'When he's locked in, they're ready. But . . .' He builds the tension slightly before continuing. '. . . crucially, he *also* needs to make sure that, once he's locked into position, no one else can butt in.' Sharon nods her head beside him. 'They must lock. They have to – otherwise other males can interfere. Rivals, in other words,' she adds.

In many insects, plugging up the female's genital cavity with a penis is a sure-fire way to stop others getting involved. That is, provided you don't get eaten during your prolonged periods of sexual incapacitation (spiders' webs are hard to escape from at the best of times, let alone while you've got a male hanging by his penis out of your rear end). However, to refer to the aedeagus as the equivalent of a mammalian penis is perhaps stretching the truth a little. It's more like a modified section of the abdomen, which contains tough (sclerotised) flaps and hooks, sometimes with snapping grabbers (known as valvae). Without the aedeagus, though, it is no exaggeration to say that understanding the wide scope of invertebrate diversity on planet Earth would be impossible, at least before recent advances in DNA analysis.

In fact, many specimen banks preserve not pinned

specimens, but the animal's aedeagus, delicately teased out of the dead insect before it withers. According to the blogger Bug Girl there are even mechanical tools for the job – namely the Phalloblaster. This handy bit of kit inflates insect genitalia with a pressurised jet of alcohol, like an industrial balloon-blower. Upon evaporating, the alcohol ensures that the tissue hardens, allowing for easier and prolonged study of the aedeagus. A niche Christmas present for any invertebrate fan.

Such is the diversity of the aedeagus in the insect world that it has spawned its own museum. The Museum of Copulatory Organs in Sydney, Australia, is the brainchild of artist Maria Cardoso. 'The sculptures twist and turn in intricate and impossible shapes. Pure white against the black surrounding them, they are beautiful and unsettling – Greek sculptures of Lovecraftian monsters,' says one popular account of her work. I guess I am sitting in Sharon and Peter's own museum.

'Would you like to come on through to next door?' Sharon asks, and I nod my head enthusiastically, for I know that this is where they do their work. We walk through. Their dining room is as regal as the living room, and as impeccably organised. Academic tomes, numerous classification keys, antique journals and collecting cases line the walls, no doubt alphabetically. On the dining table are two microscopes, along with a handful of glass wooden frames, each containing 10 or 15 shiny beetles, individually labelled and pinned with precision.

I take a seat at the table and, involuntarily, lean over the eyepieces to have a good look down the microscope. A pair of dead caddisflies gaze back at me, firmly engaged in *coitus interruptus*, if death could be considered an interruption. Each insect is about an inch long, but under the microscope their eyes look like marbles and their spongy mouthparts are both open wide in something akin to shock horror. I admire their hairy wings (which their scientific name, Trichoptera, alludes to) and their strong, hairy legs. I try to adjust my focus to see their genitalia a little better. It's surprisingly difficult. It looks essentially as if their abdomens have fused together – they are almost like one continuous body. As they lie there, bottom-to-bottom, I actually have trouble telling where one ends and the other begins.

'The one on the left is a female?' I guess. 'Nope, smaller ones are males,' says Sharon over my shoulder. I squint my eyes. 'You'll have problems seeing the genitalia, because they're so tightly hooked together.' She pulls a selection of tools toward me, a tiny metal poker, and some tweezers of various sizes. 'Feel free,' she says before popping the cups back into the kitchen. 'Feel free to what? Tear them apart?' I ask, suddenly nervous again. 'Well, feel free to have a look, I suppose,' answers Sharon from out of view. I duly pick up the smallest tweezers I can find and decide to see just how tightly held together the pair are. I peer down the microscope; the caddisflies are zombie-like and unmoving. I slide my tweezers into view and become momentarily befuddled by the fact that: a) my tweezers appear unfeasibly large under the microscope; and b) the tweezers, and my hand, appear to be shaking around all over the place. The eyepieces show nothing but a noisy blur. I pull back and check that everything is OK with my hand. It's completely still. It seems steady to me, yet I peer down the eyepieces once more and see the tweezers jiggling around again. My hands are not cut out for dissecting genitalia, I realise. In those few moments I try a number of techniques to make my hand still enough to manoeuvre the tweezers to pull the male and female apart. But it's no use. Like trying to cut your own hair in a mirror, I can't co-ordinate my hands to make the tools do the things I need them to. The two caddisflies stare back at me, mockingly – untouched, defiant. Peter comes back in and notices my ineptitude with the forceps.

'Don't worry, it comes from experience, Jules,' he says kindly. 'Learning where to rest your hand on the equipment when you're doing it, taking your time, learning to use a pair of forceps very gently so you don't squash things, or snap things, or cut things in half.' I guess the early portfolio of every professional or learned entomologist is littered with badly dissected specimens, even Peter and Sharon's. 'You can be taught most things,' he continues calmly, 'but nobody can teach you manual dexterity. You have to learn that for yourself.' Sharon walks in and, noticing that I'm a little crestfallen, chips in. 'Everyone wrecks specimens at first, or they end up flicking them across the room and losing them completely. As long as you keep all the bits of the genitalia, you're normally OK.'

Sharon and Peter each show me some of their skills. They deftly manoeuvre specimens in and out of focus, pulling, teasing, tearing – I watch silently and in awe as they reference each and every scientific name without once having to reach for one of their many reference books. Watching them at work, I can't help but feel that they may be a dying species themselves – victims of the so-called 'Death of the Taxonomist' that is often referred to in academic circles. For Sharon and Peter's knowledge will surely be lost come the DNA barcode revolution, when species identification requires nothing more than enough mush to acquire a decent sample of DNA, from which species can be determined through a computer program. I ask them about this, and they agree that the 'DNA revolution' sounds great in theory, saving time and effort and money, and yet . . . I can see that the thought of it grates on them slightly. One day, it's clear that their stock in trade – knowledge of genitalia – will sit back on that shelf in a leather-bound tome, possibly forgotten to future generations; their native dialect of insect genitalia, extinct. Forceps, needles, tweezers – these are the same tools that Darwin used; precious skills and knowledge passed down over generations from father to son, mother to daughter, teacher to student. In one hundred years' time what will we make of this knowledge? Will we laugh that once upon a time this was the best way to tell one species of insect from another? Or will we mourn those lost skills? That lost knowledge? Perhaps insect genitalia will become a gentlemanly pursuit, like those of bored Victorians learning Norse poetry, or people of the modern era that tick off Eddie Stobart lorries. I don't know, but after seeing Sharon and Peter go to work on these majestic and bizarre and diverse body parts, I feel a bit sad, in a way I had never anticipated.

Before I head off there is a knock at the door, which Peter goes to answer. It's a friend of theirs, Terry Whitaker ('lepidopterist extraordinaire', Sharon calls him). Among the many strings to Terry's bow is that he's one of only a few people on Earth studying the pyralid moths of Borneo, 30 per cent of which have apparently never been described, and nearly all of which will need their genitals given a good deal of scrutiny. Terry was 'just passing', and thought he'd pop in (I imagine that this is what life is like for entomologists; world-class moth experts like this can knock on your door, simply because they were in the area). Terry, like Sharon and Peter, is able to wield a pair of forceps like a brain surgeon, deftly pulling out genitalia or mouthparts as if he's been doing it all his life (which he probably has). The room fills with wild talk about taxonomy, rare specimens, lab technicians, pin forceps and type specimens. Not having much to offer by way of aedeagus-chat, I sit and bathe in their high passion, relishing the opportunity to see entomologists go at it like this, in their natural surroundings.

And then he's gone. Terry disappears as quickly as he appeared. We finish another coffee, and I gather up my notes and have one last look at the specimens on show, now scattered across the dining-room table. A thought strikes me, and I gather up the courage to ask. 'What drives people to want to get to know all of these complex and bizarre genital arrangements?' I ask. 'Is it pure taxonomical order that they seek, or something deeper?' Sharon chews on this for a few seconds. 'I guess I'm obsessed with them,' she offers up, honestly. 'I guess I like to do things differently. I've never gone with the crowd. I like to do things that most people aren't interested in – it's a childhood thing, I suppose. I've never followed fashion, and I used to collect things – I collected mad earrings at school, and pottery.' 'So caddisfly genitalia are your mad earrings?' I throw in. 'Yes, I guess,' and she laughs. 'I'm a little bit odd, but I guess many entomologists have this reputation among themselves for that sort of thing.' She laughs again at the thought, and Peter, smiling, leans toward me. 'Except among ourselves, of course,' he says with a quiet, knowing tone. 'Because among ourselves we're normal; it's everyone else who's odd.' He smiles and, though I don't catch it, there's a chance he winks.

My long drive home from Morecambe was one of happy inspiration, making me glad to be living in an age where I can see masters like Sharon, Peter and Terry at work in a field of expertise that can make my head happily swim. And, in a small way, that visit changed me. On my return home, for the first time in about 10 years, I retrieved my microscope from the loft. I went outside and collected four dead flies from a spider's web, brought their limp bodies upstairs, and put them one by one on slides to examine. I cast my eyes upon each, and reached for my rusty tweezers (unused since university). The first three flies I managed thoroughly to macerate, but the fourth . . . well, let's just say it was nothing if not *semi*-macerated. Something was now hanging out of that fly's bottom; if I blurred my eyes and squinted really *really* hard, I could convince myself that I had seen my first aedeagus. And how many of us can say that, huh? With every year that passes, I suspect there will be fewer and fewer. And that's a sad thing.

CHAPTER SIX

The Town that Sperm Built

'It would have been amazing if the pandas had mated naturally, but artificial insemination is the next best thing for the overall global conservation effort and the individual biology of Tian Tian, our female,' a spokesperson in the Edinburgh Zoo press release that I'm reading says. 'In the wild, female pandas will mate with several males within her 36-hour breeding window, giving her the best chance of successful conception; in the zoo this is not possible.'

It is not good news for my lady panda, Tian Tian. Once more, panda sex was a flop. Rather than risk waiting another 12 months, though, this year the zoo chose to artificially inseminate Tian Tian with Yang Guang's sperm. Desperate measures indeed.

Both pandas are doing very well and the procedures went very much to plan. After his procedure Yang Guang was up and moving within thirty minutes and back to normal within two hours; Tian Tian took slightly longer. Sunday morning saw Yang Guang back to his favourite things – eating and relaxing in his outdoor enclosure, and Tian Tian ventured out this morning.

The papers ran with it. A stock photo used on the BBC website showed Tian Tian with her head in her hands, looking crestfallen at having to be artificially inseminated. If deliberate, the photo seemed awfully ill-judged. The *Daily Mail* went with 'Birds do it, bees do it . . . but UK's only pandas need IVF as they fail to get it together to make a baby' as their online news headline ('It was AI, not IVF,' notes a member of the public in the comments section beneath).

Naively, I had never appreciated that if things weren't working out, the vets would inseminate her artificially, or even that they'd had batches of his sperm all along, ready just in case. A day later, I read that it wasn't only Yang Guang's sperm that they had used to inseminate Tian Tian. They had also inserted thawed sperm from a panda called Bao Bao, who died in Berlin Zoo last year. I had no idea such back-up plans had been prepared behind the scenes. I'd gobbled up the press releases like everyone else; the stories that sex between the two was imminent. Tian Tian reportedly getting hormonal and grumpy, him doing those horny handstands. I got caught up in the drama, in the build-up, in the sex, in the fun. And it all ended up a bit cold and business-like.

Having had only the most distant relationship with those two pandas (I had watched her bottom for no more than 15 minutes, after all), I suddenly felt immensely sorry for them both, and strangely shocked. Being totally honest, I felt a little cheated – not by them, but by the storytellers at the zoo.

I suppose if this tells us anything, though, it's that you can invest in every courtship tactic in the book (in *this* book even), but it's meaningless if sperm doesn't end up meeting egg. And that's the clincher for panda conservationists. Without that single event occurring, the species' cards are further marked with each passing year.

And so I to got to thinking about sperm and eggs – the basics of sex. I had slight misgivings when planning my 'sex tour' about including these crucial components (it felt as if they deserved their own book – written by somebody other than me), but I've caved. It took a drunken conversation with my friend, Sally Bate, to offer me a way into the topic. Sally is an equine vet. It's her fault I collared her, for she outlined to me how racehorse sex works, and I guess I had to include it. A few text messages back and forth and we were on: 'Jules! See you at 1pm on Monday for the horse-sex. Drive up the mile-long entrance, turn left and park in the courtyard. Oh, yes, of course you can bring your wife and toddler', her last text read (this decision to include my two-year-old girl may cost me dearly in future therapy sessions). Fingers crossed, then . . .

I guess first we must ask: why is horse sperm of interest to us? Well, I'd argue because it is possibly the most expensive liquid on the planet. Stallions' testicles perform alchemy, taking atoms to construct little packages of genetic codes, each packed up in little motile cells that can be sold, in their millions, for £50,000 a pop or more. Their sperm drives an industry. It powers whole economies.

Frankel was the first to set my head spinning. Frankel is a horse. A famous horse. In October 2013, he was retired from the racecourse after winning 14 consecutive races, the last being the Champion Stakes at Ascot. And he has now begun life as a stud. In his racing career he took more than £3 million in prize money, but listen to this: selling his sperm could make him (or his owners, anyway) 30 times that, while he enjoys a happy retirement, too. His has the potential to be a multi-million-dollar load. That is, if he's up to it. And many aren't.

The life of a stud is arguably no less intense than that of the racehorse. These stallions may be expected to have sex up to four times a day, with a queue of expectant mares watched by human entourages writing cheques in the background. And I've been told it's a strange world indeed: a world of teasing gates and pedestals, of unusual horse wear and handy viewing platforms from which me, my wife and my toddler were to visit and watch, wide-eyed.

Here's what happened . . .

Newmarket is a funny place. Just off the A14, far past the lost hills of Northamptonshire and further into the flatness of the Fens. The villages you pass are lovely – all of them exist in full sun in your memory, even if the day on which you visited was dark and dingy. A place where cricket matches appear to be played all year round, and posters advertise quaint theatrical productions of well-known farces. Keep driving through these villages, though, and you'll head into Newmarket. Here, farmers' fields give way to large open paddocks, and the driveways to your right and left go from being made of smashed-up bricks to gravel, to slabs, to polished cobbles, to gold. Bit by bit, fields of yellow rapeseed flowers morph into the brightest of green grass, grass that looks good enough to eat. And it is, if you're a horse. Get nearer still and you notice that, alongside every road you drive, there's another road to the left and right, each hidden by a long hedgerow. That's a road for horses, completely separate from the road on which you drive. It's a little highway for them, so they don't get scared walking to and from town. As I said, this is a strange place indeed, a place where the survival and safety of horses is of paramount importance; for each has a whopping price-tag around its neck and another wrapped proudly around its genitals.

Many regard Newmarket as the global centre of thoroughbred horseracing. It is certainly the UK's premiere place for horse-breeding, and money pours out of here across the whole local economy – it's said that one in three jobs here has something to do with horseracing, including stud farms. Country estates here aren't run by your average beagle-owning beer-guzzlers. Here, your landlords are likely to be Arab princes or the rulers of other nations.

We carry on driving, my jaw open now. After 10 minutes, we find the stud farm, drive up its mile-long toy-town driveway and park in a quaint courtyard, surrounded by all manner of new barn-conversion offices, and posh stables with blue-slate awnings. Sally gives us a wave and a broad smile as she walks over to the car.

Sally is a wonderful human being. A great vet, she's 50 per cent hard facts, 50 per cent care and understanding for both her animals and their human counterparts. Born a vet, like most vets; born in latex gloves, mucking out sheds before she could walk. You know, that sort of thing. Though she's really my wife Emma's friend (they shared a house at university), I am drawn to her when I've had a few drinks because she is one of the few people I know who can tell me what it feels like to put your hand up a cow's bottom. I corner her with a host of questions like this each time we meet. And I like to think we both get something from this relationship: I get to imagine sticking my hand up a horse's bum, and she gets to *tell* someone what's it's like to stick your hand up a horse's bum. What can I say? It just works.

We make a funny-looking group as we make our way through the compound. Me, Sally, Emma and my toddler Lettie. Everything is so clean here; no litter, spilt hay or leaf matter. It sparkles. Swallows zip up and down the courtyard, in and out of their nests in the rafters, out of view. Their chirps and gurgles are the only thing that splits the sound of our footsteps on the warm cobbles.

Sally walks us through an enormous metal door into a big padded cell with a high wall. 'This is where the mares first come when they arrive. It's passport control,' she says. 'OF

COURSE,' I think sarcastically. PASSPORT CONTROL.
FOR HORSES. OF COURSE, NORMAL. This is a
NORMAL Monday morning. We step inside. I look to my
right and notice a little booth behind a glass window in
which an official must sit and check the horse's documents
when they come in. 'They make sure it's definitely the right
horse that's coming in, and that she's got all of her documents
– including clean health certificates.' Nobody wants to pay
money to set up a sexual liaison with an STD-ridden horse,
she tells me. 'It all gets checked,' she goes on. 'If a horse isn't
clean she can't go anywhere near the next bit. If she's dirty
she has to go straight home.' I look round at my daughter
who is in her element, jumping up and down on the spongy
floor, smacking her hands on the foam walls of this, a
potentially disease-ridden animal-sex zone. I silently
commit to washing her hands when we leave. Twice.

Out of passport control, we head over to the arena. It
turns out there are two horse-sex arenas, each behind their
own enormous metal door, side by side, and each housed in
a big, tall, freshly built barn. Emma suggests that she and
Lettie find somewhere else to go while this bit goes on, but
we decide instead to stick together – anyone with any
concerns can avert their eyes if necessary at any point, or
have hands thrust over them in the case of Lettie, we agree.
Sally leads us toward a viewing platform up some stairs, and
we take our seats on some bar stools in front of the big glass
windows that look out on both arenas. The fact that there is
a plush viewing platform to watch horse sex tells you a bit
more about the financial implications of what we're about to
watch. People who visit here are so rich that they expect
executive lounges wherever they go, even while watching
their horses have sex.

Sally lays out the facts of what's about to happen. 'Once a
stallion is put out for stud, owners of mares can get in touch
with their CVs to become part of his 'studbook' – offering
up the asking price often on a 'no-foal no-fee' basis. If the

mare is deemed good enough quality, they can bring them across to the stud farm at the agreed time slot,' she outlines. The owners of these mares are putting up a lot of money for this moment, so a comfortable viewing platform like this is the least they can expect while the deed is done. Both of the arenas are the same. About the size of an 18-yard box on a football pitch (or anywhere else for that matter), the floor is made of bouncy rubber and the walls are (as with the passport control) covered by a spotless white foam padding. On one other side is a large black gate, and in the corner a small fenced-off area, with big black bars like a novelty Wild West prison.

Then all of a sudden, it begins. In the left-hand arena, on the far wall, a padded door I hadn't seen before is opened. Two men walk in with a horse. It is enormous. Sleek and muscly, it looks oiled like a Mr Universe entrant. Its muscles appear to have muscles *on them*. The horse strides in, trotting and confident. It obviously knows the routine. Then, from a nearer door, three people bring in the female, the mare. She looks a bit more nervous. She pulls back slightly, and then spins around a couple of times, dragging her tiny carers around like something out of *Gulliver's Travels*. Things from that point onward happen quite fast and I lose my grip, temporarily, on reality. All sorts of hell breaks loose in that arena. Sally kindly explains what's going on throughout the 'action', and I nod and act as if I understand, but my fragile eyes dart at a host of things that I've never seen, and will now sadly never un-see. Our stallion certainly makes a lot of noise, and does a bit of shirking and thrusting, but he also manages to fall off a few times, which surprises me a little ('he's just fannying around,' I hear someone shout from the arena). Then the stallion mounts the mare one more time, as if to prove everyone wrong; he thrusts and shudders a few times, and then suddenly everyone seems much happier. He flops off her back and is then led back to his stable, while the mare goes back home with her entourage. Done.

I've barely had time to draw breath, and we're off again. Suddenly in the right-hand arena another pair are being led in. Again, this whole thing turns out to be a bit of a blur. As is the next one. And the next. By the fifth time I've witnessed these prize horses having sex, I've gathered myself enough to work out exactly what happens here.

It goes like this. First the stallion comes in and is allowed to strut in the corner of the enclosure, as the mare is brought in. The mare's handlers lead her over to what is called the 'teasing gate', at which point the stallion and his entourage approach from the other side of the gate. They assess the mare's reaction to him approaching, and whether she's shifty or flighty (and potentially likely to kick a million-dollar stallion in the prize-bag). If ultimately she's not in the mood she is led away.

If she's ready, the stallion is then led around the teasing gate to the mare, who is by now being steadied by her handlers. The stallion, impressively, scrambles atop her and drapes his big muscular body over her back. Miraculously, and without anyone seeming to notice, he then grows a penis like a prize-winning vegetable. Not having hands, just hooves, makes this all look a bit clumsy, and it many ways it is. Horse penises appear to require a surprising amount of 'guidance' by human hands, which I wasn't expecting. The thrusting sessions that follow are enthusiastic. I notice that, while on top of her, many of the stallions give their mares a loving little nibble on the back of their head, which I'm not sure they approve of much. On the mare's face during mating is a laser-like seriousness, a mixture of terror and true grit, perhaps. It surprises me that the mares can remain quite so still and quiet throughout. And there is another surprising thing. During all of this, one man seems to have one job and one job only – to hold the base of the stallion's penis when it is in the mare's vagina. Sally informs me he is trying to feel for the tell-tale throbbing of ejaculation – without this there's no point. This man's role is crucial, apparently. He's a bit like the referee – only he can blow the final whistle. It is very interesting to witness.

But wait – I know what you might be thinking: 'Oh God, did their daughter watch all of this?' Fear not. Thinking on her feet, Emma made a playhouse using bar stools, which engrossed our little darling throughout most of the events that occurred out in the arena (Lettie pretended it was a castle, and in a way it was: a protected place that kept at bay the brutality outside).

And talking of safe-houses, I haven't yet explained what the Wild West jail in the corner was for. This was also very odd. It turned out that it was for the foals. You see, the mares often come with the kids, so to speak, so the handlers deposit these young horses in the corner, safely behind bars. It's like a kind of crèche. A crèche where you have to stand and watch your mum being ravaged by a multi-million-dollar stallion. The look on the faces of these yearling foals was not totally unlike my own as we stood up there, watching the proceedings unfold. They had a wide-eyed head-tilted look about them as they stared across the sex arena at their beloved mothers. Everyone else seemed to be wholly OK with it all, except us and the foals. After all, this is a scene that plays out up to four times daily, for a few weeks each year, across a whole town of horses. It is a sex-life that drives a whole racehorse industry.

But, fun as this is, I am here for a reason. I began this chapter talking about sperm and eggs. So what of them? It's time we had a break from horses, so please allow me to digress.

First, it's important to say that the science of sex cells is not as developed as you might imagine. For instance, the word 'reproduction' wasn't even in common use until the mid-18th century. Back then, the word 'generation' was used to describe how males and females created new life, and how organisms could seemingly grow from nothing. All sorts of theories abounded, summarised excellently in a

2012 paper written by Matthew Cobb of the University of Manchester. According to Cobb, Aristotle favoured an idea that females provided the *matter* for a growing animal's body (from their menstrual blood), and men provided the *form* through their seed. Aristotle also suggested that things without blood (like insects) spontaneously generated, but everything else (he was sure) came from eggs. Butterflies, mosquitos, clams, crabs, all of them: pop! They just turn up. That was what Aristotle thought; a deep thinker, an educated man. Hilarious really. To these early thinkers, the whole role of sperm and eggs was a fluffy cloud; a realm outside the reach of ancient science and scientific thought. But then, slowly, ideas did develop, and with them theories, some of them testable. At one point an idea came along that rather stuck. It was that semen acted like a seed and the vagina like a fertile field. Semen was the primary component, it was thought, and the vagina merely the receptacle of said semen (in fact, it was this accepted view that largely fed through into Judaism, Christianity and Islam). Like pedestals ripe for smashing, each of these beliefs was eventually to be shattered, but it was a long time coming.

It started with a Dutch scientist called Swammerdam in 1669. Unbeknown to the rest of Europe, Swammerdam's experiments revealed that insects did in fact come from eggs, not from spontaneous generation, as was widely thought. Swammerdam had the amazing insight to watch what insects *do*, rather than spout any old guff. Three years later, another Dutchman, Reinier de Graaf, put forward his observations in his book, the *New Treatise Concerning the Generative Organs of Women*. It struck gold. In a section on rabbit mating and pregnancy, de Graaf described the female rabbit's follicles reddening and rupturing, and from these he reported observing small spherical structures. Eggs. Rabbits, he deduced, made eggs. They were just hidden inside the body. Tiny unseen eggs. The same was probably true of other mammals, he hypothesised. Now this was certainly a big step forward. But none of this helped explain what the hell the semen was about. Its function took longer to get to grips with. It was involved – sure, people agreed on that – but how? No one could provide a satisfactory response. De Graaf called it merely a 'seminal vapour', and that description was one that few could challenge scientifically.

Then a new man came on the scene. His name was Antonie Philips van Leeuwenhoek, dubbed the 'Father of Microbiology'. He changed everything. Not classically trained, Leeuwenhoek is reported to have been a draper, but crucially he was a man who had access to lenses. He had used these to build an early microscope and, being blessed by scientific curiosity, he used this apparatus to scrutinise unusual things. His were the first human eyes ever to observe protists, bacteria and the inner workings of cells. And, God bless him, he had the audacity to investigate, of all things, his own semen. Now, this was the 17th century. You couldn't just pleasure yourself onto a microscope slide. No, no, no. I've heard it said that he instead obtained the semen through 'natural means', presumably by retrieving the fluid from his good wife after intercourse. He certainly knew that his research was on the cutting edge of taste and decency, though. It's likely that he ummed and ahhed over whether to make his observations public at all. In a letter to the President of the Royal Society about his findings, Leeuwenhoek insisted that if they find it 'either disgusting, or likely to seem offensive to the learned, I earnestly beg that it be regarded private and either published or suppressed as your Lordship's judgement dictates'. He was rattled. Rattled because in his semen he had seen moving things. Things he named 'spermatozoa' – 'semen animals'. We are used to the idea of sperm and eggs, but sometimes I like to imagine being Leeuwenhoek on that day he first looked at the liquid that was being cooked up in his genitals. I like to imagine him pulling back from his microscope: 'ANIMALS?'

We owe a great debt to Leeuwenhoek for being confident

enough of mind and scientific endeavour to speak up for what he saw. Spermatozoa were real, his science had proved it, but his discovery failed to answer the big question: what role did his little sperm animals play? Did they provide food for the developing eggs? Did they provide the body plans? Did sperm wake eggs up? Scientists interested in such things back then promptly divided into two camps, the 'ovists' and the 'spermists' – each arguing about which entity was doing the bulk of the reproductive work. And this argument went on for some time. It was another 150 years before scientists got their hands on some facts to settle the matter. As cell theory blossomed in the 19th century, scientists began to see that eggs and sperm were not enormously different from one another. They became simply known as sex cells. One swims. One doesn't. Simple. And they appreciated that, on the whole, both were needed to make new life.

Knowledge of these sex cells became a crucial ingredient within biology's modern synthesis, which revved up between 1936 and 1947. After this, sex cells took their illustrious position as straddlers of multiple fields of population biology, developmental biology, biochemistry, evolutionary biology, cytology, palaeontology and much else besides. From this point onwards, sperm and eggs were seen as vital genetic vehicles, capable of meeting, combining and making new, big, walking, talking vehicles, in whose genitals the process would start all over again. The reductionists loved them. Science loved them. In fact, 50 years later I was to have my own little scientific fling with them, an experience I'd like to take a few moments to describe to you now, if I may . . .

It all happened a few years ago. Sitting on a stool in my tutor's busy lab at Liverpool University, I would pop a tape of salmon sperm swimming around under a microscope into the video recorder. And I'd sit there watching it, with my notebook on my lap. Every few seconds I'd pause the video, get out my ruler and take some measurements of the sperms.

Then I'd measure speeds and acceleration of these motile little blobs. 'Did long tails make sperm swim faster?' That's the research question I was helping answer. At one point we went up to the rivers of Kielder in the north of England, to collect some more salmon sperm for the lab's research. I spent two days watching burly men 'milk' salmon into buckets, an unforgettably bizarre sight that competes for space in my brain with the mating million-pound horses with which I began this chapter (and the explosive duck penises, of course, from Chapter 4). But there was a reason for this. As well as giving us the Beatles and Wayne Rooney, Liverpool delivered a new theory to do with sperm, one that was as momentous as anything Leeuwenhoek, de Graaf or Swammerdam offered up. It was 'sperm competition', a term coined and a subject pioneered by the University of Liverpool's Geoff Parker.

Like many good stories, it started over a cowpat. Parker was a dungfly-watcher, among other things. Studying dungflies up close in the 1970s, he observed that females would often mate with one male after another after another. He deduced that, if individuals could evolve behaviours and other adaptations to maximise chances for sex, then, in theory, so could sperm. Parker contemplated how this might come about, hypothesising that in scenarios where females mate with lots of males, then adaptations for males to beef up their own sperm (in size or quantity, say) or destroy the sperm of rivals should flourish. His ideas inspired a generation of sperm scientists, the wonders of whose research we are still revelling in.

Parker's predictions proved true; generally speaking, where polyandry (female promiscuity) becomes established in a population, adaptations to counter sperm competition pop up in the males. War, on a microscopic level, is waged. Often this plays out in two ways, popularly imagined as a lottery. Males either buy more tickets (invest in making more sperm) or they rip up the tickets of their competitors (invest in adaptations that kill off the sperm of others). Some try other methods. They trash their competitors' tickets (by using sperm plugs to block the females' genital passageways), or they force the ticket machine to pay out early (through chemical 'stimulants' that trigger the production of eggs), before anyone else gets a chance. Or they purchase only the highest quality tickets money can buy (super-sperm, in other words).

Though scientists are still attempting to understand on a grand scale how such adaptations evolve, there are certainly a host of weird sperm and sex secretions out there for us to study. Olivia Judson's list of odd sperm adaptations (in her book *Dr Tatiana's Sex Advice to All Creation*) is fascinating. There are hook-shaped ones (koalas, rodents and crickets, to name but three) and flat disc-shaped ones (the protura, sometimes called soil-dwelling coneheads); there are the spinning sperm (crayfish), the corkscrew sperm (some snails), bearded sperm (some termites), crawling sperm (roundworms), not to mention the host of creatures that squirt their sperm in tandem arrangements (water beetles, millipedes, sea snails and American opossums among them). Perhaps the best-studied animal semen is that of fruit flies. Their semen is renowned for the fleet of manipulative chemicals it possesses. These secretions reduce female libido, kill the sperm of rival males and even speed up the female's reproductive cycle to get her away from the mating arena and down to the business of laying eggs more quickly. So potent are their chemical powers that it appears to take its toll on females, leading to their death at a younger average age.

Modern scientists have continued Leeuwenhoek's legacy of observing their own (and others') semen. In fact, human semen is very well studied. And many of sperm competition's calling cards are there in us, so much so that only 1–5 per cent of human ejaculate is actually made of sperm cells – the rest is a cocktail of neurotransmitters, endorphins and immunosuppressants, all present solely to combat rival sperm, enhance female mood (through a hormone called serotonin), and even induce sleepiness (through another, melatonin). Incredibly, there's even evidence suggesting that women who have unprotected sex (and therefore have regular contact with semen) are less depressed, and less likely to commit suicide. Make of that what you will.

If I could travel back in time, I would love to track down Leeuwenhoek and tell him where *his* secretions would take us. From seminal vapour to a keystone of biology's modern synthesis, in a rather mind-boggling 300 years. Oh, to show him the curly sperm of a termite, to tell him about semen's suspected anti-depressant qualities, or to describe to him the one-metre-wide spermatophoric love-bomb (containing 10 billion sperm) that belongs to the giant octopus. To tell him about those horses, and the industry spawned by their . . . spawn. To show him the lab that Sally showed me later that day, dedicated to making the stallion's sperm do incredible things. For there was plenty more to see on that stud farm in Newmarket.

'AND THIS . . .' says Sally from across the room a couple of hours later, '. . . IS AN ARTIFICAL HORSE VAGINA.' She holds across her chest and over her shoulder something that looks a little like a leather case for transporting a surface-to-air rocket-launcher. 'It's used for getting sperm samples, so that we can check up on his sperm and make sure everything's OK.' The spotless and well-lit lab is full of microscopes and other unknown (and possibly very expensive) equipment; Emma is holding Lettie's hand tightly. Sally hands me the artificial vagina. It's heavy – like an enormous leather condom. Sally informs me that its padded sides have slots in which you can pour warm water, to make it feel a bit more 'lifelike'. She smiles. 'Some like it really hot, some like it quite cool – it's a bit of a mission trying to get it right,' she says honestly. There's a little hole at the end onto which is attached a little 'collection point', a bit like a rubber glove. From here the stallion's semen is taken.

We walk over to the fridge. 'And this is where we keep the sperm extender,' she tells us. My eyes widen: bloody hell – *sperm extender*? Sally opens the door to reveal a fridge filled with cans of Coke. Momentarily I am stunned, and Emma, knowing that I am mistakenly considering whether Coke can make sperm grow, points to the back of the fridge where a row of little glass medical bottles sit. I cough. 'Sperm-extending is something we occasionally turn to in the trade; it extends the life of the horse's semen, meaning that it can hang around longer in the mare's reproductive tract, and hopefully bump into an egg at the right time.' It's a trick used on the older stallions, whose semen suffers from lower sperm numbers and poorer motility of sperm.

Something dawns on me at this point, as I stand there drowning under the weight of an artificial horse vagina. 'Why not get rid of all this ceremony out there?' I say pointing to the sex arena. 'Why not stick to a good old-fashioned rubber vagina, collect the semen and save all the bother?' It's a simple question, and one that I can't help thinking cuts to the heart of the whole racehorse-breeding industry. After all, it's cheaper, and racehorses can sire more offspring and make more money that way. Sally's response is prompt and forthright: 'Nope. You can't do it. No.' She is unimpressed. 'No one would allow it and, besides, it's illegal in thoroughbred breeding. Legally, they have to be naturally covered (vet-speak for "fertilise").'

'But, why?' I push. The answer is essentially because of the market as much as to do with holding back the profiteers. Conceptions have to be 'natural' for the sake of the future of the sport itself. Sally goes on, 'Let's say, if you allowed for artificial insemination, a top horse that might cover 150 horses in natural circumstances might then be allowed to inseminate 1,500 because they're so in demand from breeders. If you allowed 1,500 people to buy your sperm, well, the system would break down. You'd dilute the market, and you'd risk ending up with one brilliant horse's genes flooding the market. In no time at all, you'd have individuals within the sport too closely related – the sport would be ruined.' The sex arenas are here to stay, for the time being at least.

We spend the rest of the day wandering from stable to stable, feeding carrots to the boys we'd just watched have sex with the female horses, which were now miles away, back on their farms with their foals. It felt strange, like bumping into footballers in a stadium car park after the big game. Each horse snacked heartily on the carrots we offered (Waitrose, I noticed), but I was perhaps the only one who was standing there aware of the fact that some of the atoms in those carrots would find their way into that horse's semen, for use tomorrow or the day after, to fertilise another unknown mare who at that moment was standing in a field, minding her own business, miles away.

Later, saying our goodbyes to Sally, we came across a horses' graveyard. Rows of neat little marble plinths, upon which were engraved the horses' names, their lineages and their trophy-winning titles. An industry where history is written by the winners, in terms of ancestry. And genes. In that respect we're not that different: us, those horses and the pandas – all of us can be reduced to vessels for that crucial genetic explosion, the meeting of sperm and egg. Simultaneously the biggest and the smallest thing ever to have happened in the sex lives of every animal that has ever lived. The numbers game, a million times over. Life's lottery. 'You breed the best with the best, and hope for the best,' goes the old horse-breeders' adage. My fingers are crossed then for Tian Tian – has the explosion taken place? We wait. We wait.

CHAPTER SEVEN

Land of the Sexless Zombie Time-travellers

In birdbaths, water-butts, roof gutters or the cracks in the concrete on your patio floor, there lurks something that doesn't belong. It gathers on the moss that sits on your windowsill. On a dry day after a woodland walk, hundreds of their bodies lie desiccated within the grooves of your soles. Desiccated, but not dead. For this is an animal like no other, a traveller in space and time. A sexless zombie time-traveller.

The bdelloid rotifer is a creature that I've became rather fond of while researching this book. It's the *cause célèbre* of sex science because, frankly, it doesn't have sex. It just doesn't do it. Bdelloid rotifers haven't had sex in possibly 40 million years.

Now, if you haven't heard of these beasts before, you are in for a treat. If you have, then I hope to pay homage to the recent great strides that science has made in the field of sex, courtesy of these common little birdbath monsters.

Even before their modern appreciation as ancient asexuals, rotifers have always had a certain degree of popular appeal, being as they are a) weird, and b) pretty much everywhere, if you look hard enough. We've already learned about Leeuwenhoek, but a second great 18th-century microscope populariser, Henry Baker, was a keen rotifer-watcher and, being among the first humans ever to look upon them, wrote:

> *I give it for . . . distinction sake the Name of Wheeler, Wheel Insect or Animal; from its being furnished with a Pair of Instruments, which in Figure and Motion, appear much to resemble Wheels . . .* [wheels which] *seem to turn around with a considerable Degree of Velocity, by which means a pretty rapid Current of Water is brought from a great Distance to the very Mouth of the Creature, who is thereby supplied with as many little Animalcules and various Particles of Matter that the waters are furnished with.*

They are what sea anemones dream of being – looping, squirming, soarers on the wind, masters of their own destiny, and marvellously, wonderfully free from the trials of sex – or so we're often told.

My copy of the 1880 work *Ponds and Ditches* (which I surreptitiously make prominent on my desk whenever we have visitors) is furnished with many miraculous descriptions of rotifers. They can be sprightly, zipping around like a spinning-top; others just sit quietly, their cilia flickering like candles. They crawl, they loop like leeches, or shuffle like caterpillars, or drift freely across microscope slides, homing in on unknown scents. There are shielded forms, armoured forms, and some that look like ornate wine glasses. As a group they are diverse; there are erratics, conformists and every body shape in between. But it is their resilience that Victorian scientists appear to have noted with the greatest interest; their so-called *revivification* – the ability to dry and rehydrate their bodies, coming back to life, re-colonising puddles that were once parched. 'Were these animals alive, then dead, then alive? Or were they always alive?' they asked ('death . . . is not life in a torpid state, but the absence of life' is *Ponds and Ditches*'s best guess).

It's ironic, then, that a sexless creature on this planet has the potential to tell us most about sex, and why it matters so much to everything, all around us, all of the time. No sex book can be written without a chapter on rotifers. They're crucial. So I set up some meetings and headed off out to learn more about this sexless creature, one of which probably lies, desiccated, just metres from you now as you read this.

*

'So, is it an animal?' I ask. Chris holds out a knitted model of a rotifer that one of his students once made for him. It fits comfortably in the palm of his hand. Near the 'head' end sit two circles like gormless eyes (which Chris informs me are, in actual fact, gormless eyes), and underneath I am told to imagine a ciliated hole, the mouth. It looks a little, but not totally, like a knitted croissant that has been stretched long before baking. 'Animals?' Chris responds happily. 'Oh yes, fully fledged animals. They've got brains, musculature, guts, a bladder. Everything you'd expect.'

Chris Wilson is a Postdoctoral Fellow at Imperial College, London. He's one of a host of scientists who spend much of their time watching these little things whirr and crawl silently through drips of water on microscope slides. He is *exactly* the sort of person who would own a knitted rotifer.

Each sentence that comes from his mouth is perfectly packaged, each word individually checked and served up free of errors, and he handles me with the kind patience and grace of a primary-school teacher, for which I am extremely grateful (and which may explain the educational puppet).

'And here's what happens when they dehydrate.' He gently squeezes the head and the tail of the knitted character into itself, making a ball. 'Watch . . . you see that the head and tail get contracted into the body? It dehydrates like this,' he says. 'Then, when it gets hydrated and comes back to life, it just pops back out again, back into shape.' He relaxes his hand and the puppet gathers its shape once more, as before. He goes on. 'The cool thing is that they don't need to form a cyst or special egg – the whole adult animal can just contract, lose all the free water in its body, and then pop back to life when the water returns.'

I am amazed at this, having never really considered the question – what lurks in a dry birdbath, or on the dry lichens that gather on gravestones? Now I know that these places are riddled with dehydrated rotifers.

Chris's office is rather clean for a scientist. No detritus, crumpled Post-Its or Far Side cartoons blu-tacked messily to the back of the door. In the corner is a cloth-covered sofa, which faces, on the far wall, a solitary poster of the Tree of Life. On his desk sit the trademark tools of a number-cruncher, namely a couple of large monitors hooked up to his PC, and a pair of headphones to allow the hours to pass more easily. He pops his little knitted rotifer back on the desk.

'What drew you to these little animals?' I ask. Chris can't wait to answer. 'Well, they're amazingly weird organisms in themselves – but I'm most interested in the sex,' he says. 'Or rather, I'm interested in how these animals and their ancestors have managed to last so long without it.' He pauses, before adding that boilerplate scientist line I've read so often by now: 'For evolutionary biologists, sex is one of the biggest of questions.'

It's easy to forget the obvious thing about sex, and that is this: it is remarkably costly. All the running about, risking life and limb for copulation – from an evolutionary point of view, it just doesn't make sense. After all, if evolution favours those who pass on the most genes, why do we all actively seek to *halve* the genetic investment in each of our offspring? Why is sex so sticky? How has it clung on so successfully, in almost everything? It's a question that has bugged scientists for centuries. Naturally Chris has thought about this a great deal, and he gives me some background: 'In theory, if investment in males ceased, and those resources were used to instead produce clonal females, populations could grow twice as rapidly without sex,' he offers up. 'So, it's just stupid that we would waste half of our resources producing males that don't do something useful, like laying eggs,' he continues. Yet sex is almost totally universal. There's something about it that works.

It should be said at this point that some animals do dabble in life without sex. Some vertebrates – like sharks, and some lizards, including Komodo dragons – occasionally give birth to daughters that are genetically identical to their mothers (clones, in other words). Among the invertebrates, cloning in this way is more common (the list includes slugs, snails, water fleas, wasps, ostracods, thrips, bees, a host of worms and aphids). They only *dabble* in this lifestyle, though – all of them. Asexual behaviour rarely lasts for generations at a time. Most animals flirt with asexuality as the aphids do, using cloning as a means of buying more lottery tickets to the sex-show that follows later in the season.

Truly ancient asexuals – animals or plants that haven't engaged in sex for millions of years – are incredibly rare. As well as the rotifers, other contenders for such a title include a handful of ostracods and a family of mites (indeed, in 2005, one of the ostracod species thought to be asexual caused quite a stir when an article in *Nature* exposed it as being sexual after all). Asexual though the rest undoubtedly appear, they don't come close to the remarkable sexlessness of the bdelloid rotifers.

As Chris talks to me about sex's persistence in life on Earth, his eyes shine and he smiles. He clearly gets his scientific kicks gnawing on meaty questions like this. What does sex do? Why does it persist so in animals and plants? When did it start?

While some scientists view sex as a means to purge bad genes from the gene pool, Chris is more intrigued by the Red Queen hypothesis, which we've already encountered with regard to duck genitalia. This idea that evolutionary lineages are forced to keep running just to stand still may explain the prevalence of sex; sex produces new combinations of genes that keep infectious diseases – parasites, in other words – from running rife. 'According to this view, sex is a way for organisms to change the genetic locks on their defence systems whenever intruders work out how to break in,' he explains. To Chris, sex keeps individuals healthy. Lose it, and eventually the parasites will feast on the lack of variety; they'll engulf your little lineage, driving its numbers down towards extinction.

The sexless bdelloid rotifers are a gift from the Universe to him. This little microscopic animal represents an opportunity to test the Red Queen hypothesis, by asking how the rotifers have persisted asexually without being exterminated.

We haven't touched the knitted rotifer for at least five minutes, which I take as a symbol that, at last, I'm perhaps ready to go and see a bdelloid for real. We trail down well-lit corridors and enter Chris's lab, which is a little more like what I was expecting. Along the far edge is a long table that sits under a pair of skylights, illuminating two nice-looking microscopes on the desktop. Each has been set up and is ready for use, along with a few syringes and some little plastic trays in which sit slides and spent agar plates. Some of these agar plates seem to contain little pieces of moss. The walls are covered in posters plastered with written dates and various notes and comments. There, at the side of the room,

I hear the dull humming of an enormous, ageing fridge. Exactly as I imagined it would be. Chris pulls out a chair on rollers and I sit down, scared to wheel myself over to a microscope just yet. I anchor myself to the middle of the lab, idle while Chris potters about.

Behind the big fridge door I can see him poking around looking through the samples. 'Ah, this one . . .' I hear him mutter. He pulls out an agar plate with some unidentifiable detritus in it, then slides it underneath the microscope before spinning the nosepiece to x100. He sits down and whirrs his hand on the dial on the side of the microscope, up then down a few times, focusing it ready for me to peer downwards upon the slide. 'And . . .' he pauses, '. . . there.' He nods for me to come over. I ready myself for my first conscious interaction with this, one of nature's most ancient asexuals.

I peer down the microscope – 'MAGGOTS!' my internal monologue screams, and I retch slightly. I manage to translate this gargled noise into a calm rational observation which I put to Chris. 'Oh, goodness – they look rather like . . . like maggots,' I posit politely.

With maggots, it isn't so much the wriggling of the things that turns my stomach, rather it's the waves of movement that run along the body. The incessantness of it, the busyness, the marching undeadness of it all. If a body has to pulse, at the very least I want it to refrain from finding a rhythm. I want it to pause sometimes, take a breath, have a think. You just don't get that satisfaction with maggots, and I hate them for it. Looking at the bdelloid rotifers more closely I can see that I was wrong. These don't wriggle or crawl, but rather they loop – plopping their head against the ground, then pulling their tail forward, plopping their head further forward, then pulling their tail up once more – much like leeches. And they do pause sometimes, much to my pleasure – occasionally they appear to glue themselves down with their tails, lunging their head from left to right instead – foraging, I guess. In my head I expected them to be more like sea anemones, sitting silently, whirring their little cilia to pull food into their big mouths. Yet I couldn't even see the cilia of these little things beating – where were the 'wheels' everyone talks about?

Chris comes over and ups the magnification even more. I peer back into the eyepiece, and I have about three seconds to gawp at a close-up of a rotifer making its way across the field of view before it is gone. This is more like it. A glassy body containing everything one could ever need, a central line (not unlike the mercury in a thermometer) surrounded by bulbous bowls and rods. It looks like a glass-blower's cabinet in that body. In the very centre a pair of cushions grind rhythmically against one another. 'Oh my God, I can see its heart!' I exclaim. 'Well, actually that's its grinding jaws, crunching up the food,' answers Chris. Like mallets pelting seeds, rotifers smash and grind the miniature life drawn into their mouths by the cilia. Just as it passes out of view, I see them – those famous cilia – whirring around the two circles near the animal's head. The rotifer; an animated Charybdis.

Chris pops slide after slide into the microscope and we take it in turns to ogle these quirky creatures as they wriggle and loop across the slides, their cilia clearly visible after a while, once I've got my eye in. Even though Chris has done this thousands of times, he still has a glint in his eye. I catch him smiling numerous times as he has his eyes glued to the microscope, setting up the slides for me to look at.

Nice as it is to see healthy rotifers, it's the diseased ones that are of particular interest here. In 2010, Chris (then at Cornell University) started to take the bdelloid rotifers to task, infecting lab populations of them with strains of deadly fungi, drying them up, then rehydrating them after varying periods of time and seeing what happened. 'My contribution was looking at the interaction between the rotifers and their parasites,' he explains, taking the agar plate from my microscope and making his way back to the fridge. 'You shouldn't see lineages that last tens of millions of years without sex, because what happens when a parasite comes along?' I get ready to answer but realise Chris is being rhetorical. 'A parasite would run riot,' he continues. 'It would spread and the population would be decimated.' I picture the bdelloid rotifers like fields of monocultured crops in which pests are growing uncontrollably. He ignores my glassy stare and continues, 'The rotifers allow us to look at what, exactly, is going on when parasites come along, and how they escape and can survive.' He pulls himself away from the lure of the microscope for a second. 'Let me show you.'

He potters around behind the big door of the fridge, then grabs at another agar plate, which he slips back under my microscope. I move to one side as he tinkers with the focus. 'Got it,' he says before turning to me and wheeling backward on his chair. 'Now look at that,' he smiles. I wheel my chair closer.

Through the lens I see a different scene to the ones I have just been looking at. There are no looping 'animalcules', as Henry Baker described them. In the middle of the slide sits what looks like a rotifer-sized glass pillowcase, creased in the middle and trembling slightly. If I look carefully I can still make out a few of the internal organs, but most of it is stuffed with what looks like tiny grains of milky popcorn. 'Is this a . . . rotifer?'

'It's a rotifer,' says Chris gravely. 'Infected with a fungus.' I notice another couple of rotifers in the corner. They hang there motionless, like victims of a face-hugger from the film *Alien*: bags of life with no purpose other than to feed the sprawling parasitic life growing within. 'Eventually the rotifers rupture and burst, and the fungus spreads.' They look incredibly sorry for themselves, those rotifers, cringing and wibbling slightly. He shows me numerous slides like this, each like a silent battlefield. Slide after slide shows bruised and battered rotifer bodies bloated with spores and unknown gloop from infectious agents. This is what happens to populations when they dabble in life without sex. They get taken over. Without the genetic variation that sex affords, without mixing up the passwords, other life unlocks it, breaks in and runs amok. It is a scene that has probably been played out numerous times in the history of life on Earth: animals and plants toying with asexuality, being pillaged by infectious agents, then going extinct.

Chris's experiments show what happens to bdelloid rotifers when left to proliferate on their own for long periods of time. They are abused by infectious agents. Yet the magic of bdelloid rotifers is that they're still here. They persist, everywhere. After millions of years they still remain. So, what's their secret?

'It all started with theoretical work,' Chris explains. 'The thinking is rather simple. If you can just move away from your parasites, in space and time, you can keep one step ahead of them. If you emigrate to a new patch, your genes are new and unfamiliar there – you can reap the benefits of asexuality.' There's a slight dramatic pause. 'And that's what rotifers do; they move around, they dehydrate, they blow away somewhere else, away from the parasites, for a while at least.'

Essentially, in periods of drought, bdelloid rotifers outlive, and are tougher than, their parasites. Among a number of things, Chris's research has shown that the deadly fungal parasites can't hold out for as long without water as the desiccated rotifers. The rains return, they rehydrate, and then they enjoy a parasite-free window in which to flourish, at least before those parasites colonise their puddles once more. Even when they do, many of the rotifers will be off again, masters of the breeze. And they do appear to be masters. Chris focused much of his research on discovering how nimble and capable of movement the rotifers were compared to their fungal foes. He showed that in higher winds rotifers are more likely to sail off, on twigs and tiny bits of dust and leaves, than the fungal parasites.

Much of this research added weight to the theory of how such asexual lineages can persist out there in the wild world. Put simply, they colonise new places more ably than their parasites. In 2013, Chris's studies of wild rotifers (in the woodland that sits at the back of the university buildings) further supported his claims. An uncomfortable thought grabs me. 'So are there dried rotifers flying around all over the place? Are we breathing their dehydrated bodies in?' I ask quietly. He grins, when remembering his experiments. 'What amazed me, perhaps more than anything else, is that, after five weeks, when we visited our traps – essentially empty little dishes on the ground – and added water to observe the rotifers, we found them in nearly every single one. There was huge diversity.'

The asexual lines continue, thanks to the rotifers' dogged invulnerability and their flair for dispersal. Popping up here and there, over thousands of years, each bdelloid rotifer is an ancestral island. A traveller through space and time, colonising, multiplying, drying, drifting, colonising, multiplying, drying and drifting – the survivors all the while one step ahead of the parasites, like Roadrunner escaping Wile E. Coyote.

'Even up here – with the skylights open – I've occasionally seen stuff like moss and little bits of leaves blow in, and I've taken these bits and rehydrated them and sure enough, there are rotifers,' Chris tells me. I can do little else but swear under my breath in awe. Chris laughs. 'Exactly! I've now got to cover my apparatus to stop new rotifers falling into my experiments!' He lets that one sink in before adding, 'No other animals can do this to such an extreme degree as these rotifers.'

My mind swirls at this enchanting thought. I realise at that moment that I have never, in my life, knowingly stood so close to an organism that shuns sex in this way – that every encounter I've ever had with an organism has involved their recently having had sex, or at least their recent ancestors having had it. Even the bacteria that swarm in and over our bodies manage to exchange DNA, assimilating it from one another or passing it via little tubes from cell to cell.

Yet every day we brush up against these things, these ancient, flying, time-travelling asexual zombies. Biologists are used to dividing lines between 'simple life' and 'complex life' – it seemed to me in that moment that bdelloids deserved their own special label. Truly, a fine piece of work. For them, there's no chasing sexual partners, no diluting their genes, no hawks and doves, no . . . worries. They just go where the wind takes them, forever. They tour. They are the Bob Dylan of the animal kingdom.

Sure, many millions of rotifers will succumb to the dry, blowing out to sea or into the corner of your kitchen. Millions of others will be mauled by parasites, victims to the biological hackers who have long since learned all of the genetic passwords in a particularly deep puddle. But many millions survive, becoming billions once more in the new ponds and puddles and wet gutters or on our roofs or on the tops of our wheelie bins. In a world where sex is the rule, the rotifers have found a niche where the rules don't apply – it may be the only such niche ever, in the history of the planet, where such a lifestyle can persist.

My time with the rotifers is coming to an end. Chris has more important things to do than puppet demonstrations with the likes of me. We stand up, rolling our chairs back under the bench on which sit the microscopes, their slides still wriggling with life.

A thought pops into my head, and I momentarily have the confidence to play devil's advocate. 'Couldn't there be males somewhere, somehow? Males that we haven't yet seen? Males that pop up only now and then?' I imagine them turning up, like flowering bamboo, sprouting after 200 years and spreading suddenly throughout the slides. 'How do we truly know that there are no males and that they haven't been having sex all along?'

Chris seems glad I asked. 'That was actually a really big question for a while,' he says. 'For 10 or 15 years people had been trying to work out whether rotifers truly are asexual, both today and throughout their history.' As we walk down the corridor we stop briefly by his office, and he types a few words into an academic search bar that he makes pop up on the right-hand monitor. 'And you're just in time because . . .' – he clicks on a PDF – '. . . as of last week, the first sequenced genome for these things came out.'

Up pops a research paper, pitted with circular graphs and waving colours and screwball diagrams. He reads my expression. 'Obviously, I may need to digest this down for you.' Turning his head from the screen, he looks at me head-on, holding out both hands ready to use them as demonstrative tools if he should need to (it dawns on me that he is unlikely to have a knitted doll for this one).

'As you know, humans and most other animals have two copies of each chromosome, and on those sit two copies of every gene, one from your mother and one from your father.' I nod my head. 'When you have sex, the pairs of chromosomes link up, recombine and then independently assort so that you give just one copy of each chromosome to each offspring, with one of each gene. That's a process called meiosis.

'In bdelloid rotifers, though, the chromosomes don't come in pairs. Each one has different genes, in a different order.' He lets this soak in for a moment. Seconds pass. I don't nod this time. Chris tries to explain in a different way: 'With the bdelloids, you can't neatly match up one chromosome from Mum and one from Dad and give just one copy to each offspring. Sometimes one chromosome has a whole string of genes that don't occur in the same order anywhere else, so it can't be paired up. Or genes do come in pairs, but both copies are on the same chromosome.' Head tilted, I digest this. He continues, 'The strings of genes from two ancient original parents have become so tangled together that they can't pair up neatly for sex and meiosis.

Pairing doesn't matter for asexual copying, which is what they must have been doing for a long time.'

I take on board the basics. Without sex, there is no need to tidy things up into orderly pairs. The bdelloid rotifer's genome has become a child's messy bedroom: stray garments, unpaired socks and miscellaneous hats and gloves shoved back into any and every drawer.

'Turns out it's even more interesting, though.' He pulls another research paper up on the screen. 'The other really weird thing about the genome of these little critters is . . .' – he does some more tapping on his keyboard to build up the tension – '. . . they seem to have managed, somewhere along in their history, to have picked up bits and pieces of DNA, not just from other animals, but from other *kingdoms of life*.' 'KINGDOMS?' I almost shout it. Chris nods enthusiastically. 'When the genome was sequenced,' he continues, 'they found all these bits and pieces of genes that didn't seem to fit with other animals, but instead seemed to be bacterial, or fungal, or plant, in origin.' He looks over at life's phylogenetic tree, blu-tacked onto the far wall. 'What seems to have happened is that at points in their history they have borrowed bits of DNA from all over the place, probably from whatever they were eating. And eight per cent of their genome comes from these other things.'

In my head, my vision of the bdelloid rotifers as prudish is shattered. I suddenly picture them ransacking the rest of the kingdoms of life, pilfering hunks of DNA from other creatures like sailors on shore-leave (but without the sex). These ancient sexless zombie vampires, travellers through space and time. Bizarre, to say the least.

'What does this mean, this stealing DNA business?' I ask. 'Could this be part of their suite of skills needed to survive without sex?'

'Who knows – it seems it should be part of the story,' he nods, looking almost wistful, 'but it has very different consequences from sex as we know it.'

These little rotifers have many secrets yet to tell, and people like Chris will be among the first to know. I envy him that – he is one of a long chain of scientists besotted by these ancient creatures, from Henry Baker in the 18th century looking through those early microscopes, to today, when the bdelloid rotifer has become a star, a species around which international conferences are held, a test case for how animals live without sex, a habit to which the rest of us seem forever bound. And yet, over the centuries, these scientists – then and now – are studying the same organism – literally. Clone after clone after clone of them. The same clones now as then.

Before I sign out of Chris's building and hand back my security badge, he gives me a friendly handshake and wishes me good luck before heading chirpily back to his rotifers. He has in him the same zest for microscopic animals – the same wide-eyed reverence – as those Victorian scientists with which I opened this chapter. In fact, since visiting him I've come across a quote (from Charles Kingsley's 1859 book *Glaucus, or the Wonders of the Shore*) that seems as applicable now as it was 150 years ago, and reminds me fondly of my day spent in his company.

> . . . *no branch of science has more utterly confounded the wisdom of the wise, shattered to pieces systems and theories, and the idolatry of arbitrary names, and taught men to be silent while his Maker speaks, than this apparent pedantry of zoo-phylology, in which our old distinction of 'animal', 'vegetable' and 'mineral' are trembling in the balance, seemingly ready to vanish like their fellows – 'the four elements' of fire, earth, air and water.*

I can still see them now, weeks later, in my mind's eye. I can even see their maggot-like movements, and the circular frame of the microscope lens around them. The anti-romantics. It's been two centuries, and still the rotifers are testing the strength of those central pillars of scientific knowledge; shedding light on the wonders and reasons behind the evolution of sex, and an explanation for why, oh why, animals find it so hard to shake. 'What is sex for?' wondered generations of scientists. Now, truly, we may be closer to knowing. And who'd have thought that the answer, my friends, would be blowing in the wind.

CHAPTER EIGHT

The Human Frequency

It's been a rough few months. Between May and July each year I book myself out five days a week to schools around the country, offering myself as a sort of visiting pond-dipping man. I turn up with my net and work with the teachers and the pupils around the pond, pointing out what's what and helping them identify what lies beneath. Then I go home, scrub my hands clean, wash my hair, and do it all over again the following day.

What is it about ponds? I love them. I have always loved them. Is it the horror that excites me? The thrusting dragonfly jaws? The scything mandibles of the great diving beetle larvae? The bloodworms and the leeches? Maybe. But also it's the accessibility of this underwater world; that we can have such wonders in our backyards or our schools, or in our nature reserves, without having to hire a boat or wear scuba gear. That all we need is a net and a tray and then this wonderful world is ours to explore.

My work at this time of year involves early starts (usually before 5 am), but frankly, I love it. It's the only time I will respond to such names as 'frog-man', 'pond-man' and occasionally 'duck-man' (schools often seem to use this strange binomial naming system when having visitors – I have no idea why).

This year has been different, though. My head is swirling with information about sex, and on multiple occasions I've found myself coming dangerously close to exposing, in full gore, the rampant sex lives of pond animals, while chatting to small innocent children. Inadvertently I might teeter on the edge, and then suddenly catch myself. If a child asks why a big water hog louse is carrying a smaller one underneath I might say, 'The big one's a boy – a male – and he's protecting a little female, a girl.' This is true, but I refrain from adding that he's protecting the female from sexual encounters with other males, rather than from vengeful predators. Likewise, when we watch damselflies darting across the water paired-up, and we 'Ohh' and 'Ahhhh' when she pops her long abdomen hurriedly into the water to deliver eggs below the surface, I hold back from telling them that the male has just freshly scraped her reproductive passages clean in a way that Sharon and Peter Flint explained so eloquently to me a few weeks ago. I have become a master at giving a little, for educational purposes, but not too much.

Here's the thing, though. When you do simply tell a five-, six- or seven-year-old about sex in animals, they generally just accept it. 'To lay her eggs, the male and the female need to pair up – we call that 'mating', I'll tell them. And they nod their heads and look more closely. Later, I'll overhear them say to their teachers, 'Miss, those back-swimmers are mating!' Sometimes I hear them quietly repeat the sentence to themselves ('the water boatmen are mating') again and again. I assume that it's the first thing they'll tell their parents later that day, and I guess I'm OK with that. It's really quite lovely. It's good to hear them using such language fresh from connotation, unsullied by our notions of smut, and free from eye-rolls or tuts. This is basic animal science, after all.

Yet there are some stories that I do find myself holding back from telling. The water boatman, for instance, which calls out to females by scraping its penis against a little fold of skin about the width of your hair, creating the loudest animal sound (for its size) on Earth. 'You see that one there . . .' I want to say in my most enthusiastic voice, '. . . it's using its PENIS to blast out a noise so loud it could make your ears bleed! BLEED!' But no, hold it in, Jules – you'd never be invited back.

Whirligig beetles allow for much more accepting conversation. You can stare at these little beetles for hours with children and get right into some pretty technical information, without once mentioning aedeagi or multiply-mated females. They are shiny, oval beetles, about the same size as your little fingernail. They gyrate in feverish looping circles across the pond surface in spring and summer. Watch for a while and it becomes a hypnotic spectacle – they dash and loop, making hundreds of tiny trails on the surface of the pond that then quickly dissipate. But it's not random. They're thinking (to a degree), and often about sex. The positions that males and females take in these swirling masses are the subject of rigorous study, and it's likely that each little beetle is weighing up where's best to be in terms of staying safe, finding food, and securing mating opportunities.

Their aggregations remind me of a sexually charged roller-disco. Like bats, they undertake their manoeuvres through echolocation – reading (with specialised antennae) the sound waves that bounce back off the objects nearby, including members of the opposite sex. If you scare them (and I offer you no advice on how to do this) you might notice the females seek shelter in the centre of the swirling mass, while the males stand firm on the edge like bouncers, guarding this time against predators, rather than one another. What is perhaps most impressive is the co-ordination as they swirl around. They're like bumper cars that never crash. They're always calculating, alert, aware – even while sexed-up – making what must be hundreds of calculations each second, judging speed and distance, all the while observing visually the events above and below the surface; like archerfish, their eyes are split along the horizontal axis, allowing for both terrestrial and underwater vision. And each is just a tiny beetle. What could be more humbling to a child than knowing that a beetle can compute such complex trigonometry? Like most beetles, they deserve more respect than we might offer.

But my pond season is over for this year. It is the summer holidays, which means that there will be no more early-morning starts for me. No more having to get into bed early. Finally, I can get out there, stay up late, and see some night-time sex activity without worrying about how I'll feel in the morning. For there is one animal I have been desperate to tick off. After 33 years, it is time for me to see glow-worms – one of nature's 'must-see' spectacles.

But there is more to my quest than that. For not only is this a chapter about how boy meets girl, it's also a chapter about how boy flies straight past girl, and chooses instead to try to have sex with a lamp-post. Such is modern life, and the all-encroaching reach of ours, the human frequency.

*

10.05 pm. I am sitting on the curb in a pub car park, making eyes at a man standing by his car fifty yards away. He's definitely my man, or one of them. Karrimore jacket, waterproof cap, torch, a tough sturdy pair of walkers on. He looks up quickly at me, squints, then busies himself at his car again. Awkward.

Many guided nature walks start this way. Wildlife NGOs think it's so easy when they organise guided walks for the public. 'Meet at the Robin Hood pub car park!' they say in their adverts. '10 pm!' Well, that's all well and good, but it's dark at 10.05 pm and I feel a bit weird wandering up to people outside a pub, bumbling about like Hugh Grant, asking strangers, 'Excuse me, are you guys here for the glow-worms?' ('You what, mate?'). The smokers outside the pub looked understandably confused at my interruption of their otherwise totally nature-free lives. Is glow-worm a pseudonym for drugs? I panic. Maybe it is? Deep breaths, Jules. I wait a few more minutes, standing a bit nearer to my safe place, my car. I'm probably too early. That's it . . . too early. Then I notice them. Two women sitting under the bright car-park lights on two decorative rocks in front of the pub. Both wear good sturdy walking boots like my man over there. There are clipboards. One has a head-torch. Got to be them, I think. Got to be. I wander over and offer a comedic 'glow-worms?', at which point I notice about 30 people standing in the street opposite, all of whom are wearing good sturdy walking boots too. The two ladies have a cheap laugh at my ineptitude.

One introduces herself as Anita, our Wildlife Trust guide. She has a kind manner, professional and friendly, but there might be a hint of concern in her eyes at being responsible for the safety – out on a nature reserve in the dark – of 31 people, nearly all of whom, like me, don't have torches.

I give my name, apologise (as is my way), and cross the road to the rabble of fellow glow-worm virgins. David Seilly, our glow-worm expert and guide, is addressing the crowd. Already he weathers a polite rain of questions from the other attendees. 'How many might we see?' 'Will I need my waterproofs?' 'Will there be toilets?' The usual. I see a familiar demographic – smiling, friendly, middle-class, grey-haired people. Some of them are holding hands. Essentially, they are everything I hope to become in later life: healthy and happy. A handful of students are also present, bringing down the mean age by all of two or three months.

I have never seen glow-worms before. I'm excited, but I try not to set my expectations too high. I quash any thoughts of the scrub being lit up like a Christmas tree by their little bodies, and instead imagine seeing only one, like a single Christmas tree bulb shining from deep within the bracken. The thought of seeing even one still excites me, though. I love what they stand for. An animal that throws caution to the wind, screaming not through feathers, or through squawks, songs or dances, but through the medium of photons pumped out of its backside. 'Come to the light, baby,' she says gently to the males. 'Come to momma.' It all sounds so preposterously *not* British, somehow. Too bizarre, too tropical to find such creatures here where the summer rains are rarely warm and refreshing and are instead all too often dank, streaky and creeping.

Anita ticks off our names one by one on her clipboard, lists out the health-and-safety bullet points, and off we go, making a long merry chain to our venue for the night, Cherry Hinton Chalk Pit, managed by the Wildlife Trust for Bedfordshire, Cambridgeshire and Northamptonshire. According to the Wildlife Trust website, the former quarry provided the hard chalk to build the colleges of Cambridge University up the road, and the lime for the cement. East Pit, where we're headed, was quarried right up until the early 1980s. According to Anita it is a big bowl of exposed chalk, now scrubby and riddled with more than 60 bird species, some unusual plants (don't ask) and, of course, those glow-worms.

After we move through the gates David, our expert, stops and, without anyone saying a word, we form a compact circle around him. The summer sun has been down for almost an hour, and we are almost totally invisible to one another, except for the rumbling orange glow of Cambridge's streetlights, reflecting onto our bodies via the thick clouds above us. Our eyes adjust to the restricted wavelengths. David is perhaps 40. He has short white hair and a soothing voice. A really soothing voice. He sounds a little like a radio DJ – rolling sentences, with a rhythm that could sail through dead air. A likeable expert through and through, David has the endearing habit of being quick to underline the areas of his wide-ranging knowledge that aren't backed up by reams of scientific papers.

Before he has time to explain much, the familiar light rain of questions begins once more. In fact, what starts out as a bit of simple chat has turned into a full-on Q&A. 'Do they have predators?' pipes up a man with beard. 'How do they overwinter?' asks a woman with a stick. 'At what stage do they pupate?' asks a foreign student from the back. David responds to each question succinctly and within a 30-second time frame, like a well-rehearsed TV expert (he probably is). 'Excuse me,' says a quiet voice to my right. 'But what *is* a glow-worm?' I think a few of us visibly exhale. 'The glow-worm is a beetle,' responds our sage. 'We've only got one species in Britain. They belong to the firefly family. This one's strategy is rather simple: large flightless females emit light, with the aim of attracting males in order to produce the maximum number of eggs.'

Suddenly there is a moment of calm. The traffic behind the trees is momentarily quiet. No one says anything further.

'So, shall we go and see some?' offers up Anita. There is a positive mumbling of middle-class approval. We venture forth. Onward. As we trek, we form a great chain down onto the nature reserve path, through the trees and then suddenly outwards into the desolate bowl that is East Pit. Now fully used to the dark, my eyes take it all in. It looks magnificent bathed in orange, lit as if by a huge electric grill. The whole place is about twice the size of an Olympic stadium, with tiers from AA to ZZ that loom from all sides around us. Its paths glow in front of us like looping athletics tracks; they swirl and circle around and in front of us. Warmth seems almost to emanate from out of the chalk, radiating into the already humid summer night.

I had imagined that the next bit would take some time. That like all good nature-writing stories, we would search and search and then search some more, and then, just as we were packing up, we'd see one: glowing like a beacon, a single revolutionary invertebrate in a world lit by artificial coal-fired lights. Our glow-worm. We'd whoop for joy, hug, weep with the wonder of it all. But no. It wasn't quite like that. It wasn't like that at all. Instead it was rather more simple, unnervingly so.

It happened like this. We turned a corner, looked at the first long bank of vegetation and saw them, five or six twinkling stars in the grass. And that's the first thing you need to know about looking for glow-worms – it's remarkably easy. You don't have to get your eye in as you do with so many sorts of encounters with nature. No, look for little points of light, then, well . . . walk towards them. And that's it. Walk towards them. They look rather like green fag-ends from a distance, deposited from an unthinking smoker's speeding car ('Nature's fag-ends' – there's a tag-line for them).

Within what seems like seconds, small clusters of five or six people bend over, or sit or stand excitedly around, each ghostly green point of light. I head toward the nearest. Down low to the ground like this it's hard to make out the faces of the others in the group, but their wonder is palpable. 'My God, will you look at thaaaat,' says Unknown Man in Dark #1. 'Isn't that just amazing?' says Unknown Man in Dark #2. 'Incredible,' says Foreign Student in Dark. 'You can see that they're slightly grubby like things,' says a slightly well-to-do lady (who apparently knows this place well) as someone shines a torch right into the beetle's appendage-laden face. Even with the full beam on her, the glow-worm continues pumping out her green charge, illuminating the strands of bird's-foot trefoil on which she clings. She is long and thin, like an elongated woodlouse, and about three times the size. Her long tapering tail is waggled over to one side, and it is from the final three segments that the ghostly glow emanates. 'Will you look at thaaaaat . . .' says Man #1 again. Man in Dark #2 gently scoops the glow-worm from the floor and places her on the flat of his hand. The green glow illuminates the deep wrinkles of his palm. We all let out our own little noises of wonder.

'How does she make this light?' asks the foreign student. Man #2 seems to know his stuff. 'Luciferin and luciferinase, and these chemicals are oxidised by ATP, I think, producing the energy – she can turn it on and off too.' At that moment, the female glow-worm chooses to dim her light a little bit. 'Perhaps it's time to pop this one back,' says Man #2. He tilts his hand a little too quickly and the glow-worm falls, then disappears into the long grass at our feet. With her suddenly gone my immediate reaction is to check she's not in my hair or scuttling along my shoulders, but no, the familiar green glow shines once more through the scattered blades of grass by our feet. Even losing the bloody things is difficult.

We amble in our own little clusters from this point on, our amorphous glowing orange heads and hands the only things visible in the reflected street-glow. Each of us, in our groups, homing in on more tiny glowing green bottoms among the undergrowth. Like whirligig beetles we move. Ten, twenty, thirty – our noises of wonder get no less embarrassing. Minutes, then hours, it seems that we spend, moving from glowing speck to glowing speck. 'My God, will you look at that . . .' Man in Dark #1's words echo across the dusty bowl.

I don't want to use the following term, really I don't, but I'm going to throw caution to the wind on this: it was magical. There, I said it. Magical. It's no wonder that literary types have been so drawn to glow-worms. Ghostly, easy on the eye – a kind of low light that says 'we come in peace' to humanity. It's their public image that surprises me most. A beast that lights up on its own? You'd think our flighty ancestors would have labelled them a trick from the devil, wouldn't you? Instead, we have them up there on a special pedestal: warm, friendly beings worthy of sharing a bed and snuggling down with. In literature, references to them are positively cosy: Pliny the Elder ('glittering stars'), Shakespeare (their 'pale ineffectual fire'), Wordsworth (an 'Earth-born star'), Thomas Lovell Beddoes ('our still companion of the dew . . . with his drop of moonlight') and Samuel Taylor Coleridge (the 'love torch'). Churchill said, brilliantly, 'We are all worms. But I believe I am a glow-worm.'

They are the People's Princess of beetles, then. But still . . . something's not right. I'm not totally sold. In all honesty, I'm left a little sore that the whirligigs or the water boatmen, both of which use similarly astounding tricks to entice the opposite sex, fail to capture our hearts in the same way as these funny little beetles. We are such suckers for fireworks.

I find myself near David, and listen for a few minutes to him answering more questions from his growing flock of admirers. 'And were they once everywhere in Britain?' asks someone to my left. David chews on this for a second. 'I get the impression that they're probably confined to a few strongholds now, yes, and . . . yes . . . they were once probably everywhere,' he says, before mulling over a thought in his head. 'I'd quite like to contemplate the impact on male glow-worms of all of the streetlights,' he adds quietly, looking at the reflected streetlights bouncing back off the clouds above. Only a handful of us seem to hear it. 'Wait, what?' I bark. We all seem to stop at the same moment. Those who heard him say it stand silent there, thinking about this for a few seconds. WHAT? In the orange haze I can almost make out smiles on some of our faces. 'Streetlights?' someone from behind me says. David considers his answer and then seems to say to hell with it.

'The trouble is, the males probably go and mate with the streetlights rather than mate with the females.' They do *what*? There is a titter from the crowd. 'Is that why they're declining?' someone asks to my right. 'Well, it might be,' offers David.

I stood there and imagined what a streetlight must look like to a male glow-worm. Impossibly long strip, sultry red tone, that irresistible sexy hum. The males drawn in on tractor-beams, flying straight past the females. I picture their final hours, bashing-bashing-bashing against a panel of illuminated glass until they expire or succumb to a passing bat. Poor little suckers. It was the first time I had ever really consciously imagined that human actions, human insight, human ingenuity, human technology, could mess up the sex life of another animal. I play with the idea of the females developing an obscure body dysmorphia, forever trying to live up to that male glow-worm ideal to which they may never evolve; the long, thin, orange strip-light ('curse my dull green glow!'). The thought grips me for a moment and without thinking, I clear my throat and I say quietly and in wonder: 'Our lights . . . our torches . . . we must look so *sexy*.' There is some nervous laughter (we are, after all, alone, late at night, in a strange place and no one is quite sure who I am and whether I might be an axe-murderer). It certainly got me thinking, though. What other sex lives are we inadvertently screwing up with our human pursuits? How else is the human frequency – the buzz, the brightness, the noise – affecting the sex lives of the animals around us? Are we civilised folks becoming nature's cold shower? Predictably, the answer is yes. And not only are some aspects of nature's sex taking a battering; in some cases, it might actually be fighting back – modifying its advertising to be better heard over humanity's din.

Roads are one such battleground. It appears that here, with cars acting like super-predators picking off prey, natural selection is working at a rapid rate. And beside these roads, grasshoppers are becoming the study species of choice for this sort of thing. At least some are adapting to the ruckus.

Grasshoppers make their calls by scraping rows of tiny pegs on their back legs against a thickened vein on the forewing. Each grasshopper species has its own call, simply determined by the number of tiny pegs and the rate of this 'stridulation'. Because grasshoppers often share habitats with a number of other grasshopper species, natural selection has driven each to stand out from the others when calling. When comparing populations of bow-winged grasshoppers from locations near busy roads with those away from roads, some German scientists in 2012 spotted some key differences. They found that some grasshoppers from noisy habitats try to boost the low-frequency parts of their song, to get their voices better heard against the low-frequency drone of traffic. Nature is fighting back, and the low frequencies are the battle-lines. In fact, the low-frequency parts of other animal songs appear to be similarly vulnerable to the blaring noise of road traffic. In early 2013, a Canadian study showed that the presence of lower-frequency elements in a song could be used to predict, to a degree, the abundance of songbirds setting up shop near a road. In human terms, it seems that low-frequency singers lose patience with roads and think, 'screw this: I'm off somewhere quieter.'

Such results are fascinating, not least because they highlight how animals like songbirds may be unconsciously calculating the effectiveness of their efforts, and looking for the best time, and the best place, to let rip with a song (some research suggests that this is one reason why birds sing first thing in the morning – after all, sound travels further on a cool morning).

Similar such studies have been undertaken, comparing noisy and non-noisy habitats and looking at the birds that sing there. They hint at the same thing: lower-frequency sex songs are being drowned out by the human din. In the

Netherlands it's observable in great tits, many of which have more high-pitched songs in towns and cities than their countryside neighbours. In Germany, it's the nightingales. They sing up to 14 decibels louder nearer roads than in nearby forests. And San Francisco's sparrows are also more chirpy than they once were, particularly in the higher registers. Here three 'dialects' once flourished among sparrows. Now only one dominates: the most shrill and easy to hear over the rumble of traffic.

What's most impressive is that animals are adapting, and relatively quickly. What's unclear at the moment is how exactly they manage this. It could be that, somehow, they are listening and responding to the surrounding lower-level frequencies. Or it could be that a genetic shift in song behaviour is occurring within the gene pool, because some of the population aren't heard and simply die off unmated, their genes lost. Either way, the dawn chorus is losing its tenors.

Though less about it is understood, another place where the chorus may be changing is underwater. Here a host of animals, including fish, whales, dolphins and even invertebrates, depend on sound. Some use it for hunting, others to detect predators or prey, but many use it for sex. It may be that they are being drowned out, too. Sure, cars are noisy, but have you ever heard the racket an oil-company ship makes, one towing air guns that fire fusillades loud enough to detect the bounce-backs of oil reserves under the rocks? No, neither have I, but I suspect Flipper could tell us more if we taught him the sign for 'THAT IS A VERY LOUD NOISE MAKE IT STOP'. The same goes for those undersea construction operations that drive piles into the sea floor, which they then explode. These noises travel for hundreds of miles, perhaps more. To sea creatures, we may be the neighbours from hell.

What most concerns some people, though, is the rapidness of the increase in sea noise: in some places there may have been a hundred-fold increase in such noise since the 1960s alone. Mind-boggling, really. And worrying. Whales and dolphins could provide a useful model for research into such impacts, largely because their sounds are relatively easy to study, being as they are among the loudest noises any animal has ever made (reaching as much as 188 decibels in the blue whale; that's only a little less loud than strapping a grenade to your head and pulling the pin). These calls can travel more than 600 miles, which is equivalent to the blue whale hanging from the ceiling at the Natural History Museum in London having a chat with the model blue whale in the Natural History Museum in Gothenburg, Sweden.

The purpose of these calls is still being debated. They are likely to communicate a number of pieces of information, including species, activities, location, social calls and, of course, sexiness. Could the calls be affected by ocean noise? The jury is still out, but it's becoming a hotly debated topic. Many agree that increasing ocean noise is likely to be affecting their lives and loves, at least a little. For some species it might mean nothing; they just shout louder. But for others? It's an area of research that might provide fascinating insights in the next few years.

So what of streetlights? If they have potential to affect my new sweetheart, the glow-worm, could they affect other such life? It's well known that moths die through attraction to artificial lights (though the impact this has on populations as a whole is unknown). If serious and true, a decline in moths could have an impact on the sex lives of plants that flower only at night-time. If the moths are too busy hanging around the lights, and not pollinating the flowers, it might be lights out for both, evolutionarily speaking (moths have far bigger problems than this though, of course – habitat loss and fragmentation being key issues).

Though it's unlikely that they kill off whole populations, streetlights undoubtedly have the power to change invertebrate communities, possibly some even for the better.

Recent research suggests that the ground beneath a newly installed streetlight can become an attractive place for predators and scavenging invertebrates like ants, harvestmen, amphipods and ground beetles – plenty of food and, perhaps, plenty of sex for some as a result. But, as in the seas, it's early days; there is much more research to be done, and maybe such concern is overstated. What's surprising is that, as with marine life, so little research is being done on the impact to invertebrates that night lights might have. After all, it's estimated that the use of artificial lights increases at a rate of 6 per cent globally each year. Might this be something we regret not studying sooner? My point is that these are known unknowns, as Donald Rumsfeld once said, and I hope that one day that they will become fully fledged 'knowns'.

But I digress. It's time to get back to those glow-worms.

I check my watch: 12.30 am. Time to leave Cherry Hinton Chalk Pit, and get on the road back home and to bed. Of the hordes of glow-worms we've seen, each has been surrounded and intensely scrutinised (and even at the end, Man in Dark #1 is still offering up an occasional 'my God, will you look at that . . .').

I offer my thanks to Anita, the Wildlife Trust Officer in charge, who points out that it's a shame we haven't seen any males. Cripes, males? I had spent so much time worshipping at Luciferase's altar that I'd forgotten all about them. Anita describes them as long, chunky beetles with a head upon which sits a wide-brimmed crash helmet (I later see pictures of them and she's spot on: they look like click beetles wearing safety helmets). So splendid were the females' fireworks that I hadn't even thought about the males. I say goodbye and wander back, alone, through the silent orange-dyed streets to the car. When I close my eyes I can still see them. Their green dots shine across my mind's eye like pierced holes through black card. 'My God, will you look at thaaat . . .' Man in Dark #1's words trample upon my inner monologue.

I recommend that you to go and see glow-worms; they are one major highlight of my sex tour so far. The Wildlife Trust offers numerous walks to see them on many of their glow-worm sites across the UK, usually throughout June and July. Take your children if you have them (and see their little faces light up).

It's 1 am by the time I pull out of the pub car park. I'm tired. It's time for home. The road that should take me home, the A14, is closed, so I detour through the minor roads. Even now, at this late hour, the thermometer on my dashboard says 21°C. My headlights illuminate the bodies of thousands of moths, busily foraging for food and sex, measuring concentrations of pheromones like robot Geiger counters. I breathe deeply, trying to salvage any distinct smells, any of the chemical trails that may lead, for them, to sex. For a few moments, I try to free myself from this, my human frequency. I open the window and smell the air again. Nothing. My attempt is futile. My car stinks of pungent pond-water. I try again. There's a whiff of stale feet from the footwell.

Like marine snow lit by the headlights of a deep-sea submarine, I smash through hundreds of moths on that tiresome journey home. Some as big as ghostly bats. Some little more than whirling bits of belly-button lint. They waft up in front of me like steam off the road. Dozens of them, bashing into my bumper or hammering against the wing mirrors. Some ricochet, or sound as if they might chip the glass. I wince each time, guilty for their stolen sex. Their animal frequency – interrupted by my own.

CHAPTER NINE

The Insurmountable Hump

We pull up to our holiday spot, an hour south of Aberystwyth, in the late afternoon. It's a farm within which sit three converted holiday cottages, one of which will be our home for the next week. On arrival, a border collie leads us down the farm drive, ensuring we stop for the ducks and the chickens. Butterflies flapping, red kites mewing, horseflies sunning themselves on the bonnet – summer's all present and correct. Idyllic, really, or pretty much. Little do we suspect the gruelling hour or so that is to come.

We park up and Steve, the park owner, comes over, introducing the collie as Billie – his trusty go-to and sturdy sidekick. We shake hands (with Steve) and he gives his wife a holler from the kitchen. She sees us and waves through the window. And then he mentions their little one, who spies us from across the farmyard and struts confidently toward us. He's still a babe, but cocky. So cocky. So cocky that I can't bear the thought of him getting anywhere near my toddler, who's still strapped in tightly, in the back seat of the car. Their little one is bloated. Bloated on a sort of confidence one rarely sees. He strides straight up to me, and I try to make out I'm OK with the fact that my personal space is likely to be quickly intruded. He lunges at me with his snout, mouth open. Shocked, I momentarily lose my balance and fall. 'Try not to let him do that,' says Steve about this, his pet pig.

It might surprise you to know that I have never, in my whole life, been face to face with a pig before, let alone a brash young pig like this. I had no idea that they can be so downright brutish. I look down at my leg and notice an imprint of his muddy snout on my jeans, like a hot slap left shining on my cheek. He snorts around in the mud underneath my feet, pushing my legs left and right, like I'm nothing. My calves rub against the coarse white hair of his back as he weighs up the calorific content of my flip-flops. 'This is Olly, our pig,' says Steve, brimming with pride at what is clearly their golden boy.

I try to make out that I think it's great. I'm with it. I love animals, remember? Except, well, I'm not OK with this. I pretend I am, but I'm not. At six months old Olly is already about the size of a boiler, with eyes like demonic pilot lights and nostrils like dials. His face looks like the split end of a cheap barbecue sausage. And he is so supremely confident. About everything, really. I mean nothing to him; worthless, unless he can find some way to get through my skin to the marrow-filled skeleton beneath. I refrain from mentioning this to Steve, though. Instead, I choose to lie profusely through my teeth as I pat Olly's big bottom. 'He's a funny chap, eh?' I laugh. 'You looking for a snack?' I offer up cosily. My wife and daughter have stayed in the car with the doors locked. Olly has wandered off and is now tipping over the recycling bins behind the car, biting the end off an empty glass cider bottle. 'Olly, no,' say Steve witheringly. He runs toward him. 'NO!' He pulls him back by his left leg and gives him a little jab, which Olly barely notices.

A few moments later Olly waddles around the front of the car just as Emma plucks up the courage to get out. She sees him approach and gets back in, nervously. He starts snuffling up the lawn, laying ruin to the decor. 'Olly, NO. NO, OLLY.' Steve comes up behind Olly again, puts one leg either side of his body and tries to steer the pig away from the lawn and toward something else. Anything else. Olly decides that all along it was gravel he wanted, and he plunges his heavy face right into the task: mouthing, sniffing, scoffing and spitting out stones as if his life depended on it. This is the point at which I really take umbrage. He is a pig. A well-fed pet pig. Why eat gravel? Why?

I have never seen an animal behave like this, and I've been watching animals for pretty much my whole life. Emma and Lettie peer tentatively through the window. I'm trying to make out to my daughter especially that this is nothing to worry about, and that she's lucky that she might be coming face to face with a real-life pig. 'Look, a pig!' I give her an encouraging nod that says: 'JUST LIKE PEPPA!' I try to hide the snout-print on my leg. Lettie can tell from my eyes that this is just an elaborate trick.

Eventually they do somehow manage to sneak out of the car, while I make small talk with Steve. In the space of 10 minutes, Olly has tipped over a plant pot, cracking its base ('Olly, no, we don't do that!'), chewed a watering can ('NO!'), and tried to maim Billie the collie ('Billie! NO! Bad dog!'). Before we unpack, Steve suggests we go for a walk with him around the farm to see the rest of the animals. I barely hear him. 'You're thinking about the pig, aren't you?' he says quietly. I nod sheepishly.

We meet Steve and the other tenants (all equally confused by this strange ritual) in the farm courtyard a little later. Our little girl is sitting on my shoulders. Her fingers pull at my throat as we hear snuffling and the sound of trotting across the courtyard. Olly. Of course, it's Olly. It's only been minutes, but he seems to have grown. So chunky is he now that he moves like a scrum; forward motion seems to take the straining power of an army of groaning men, each with that same determination to progress, to struggle onwards, every step a marking post. 'Ah, Olly,' says Steve gladly as Olly smashes into us all, assessing the order in which he'd make paste from our flesh. Emma looks at Lettie on my shoulders, checking that she is securely in place, and wondering whether there might be space for one more. We stand there for a few moments, there in that courtyard. Steve gives the pig a long playful scratch on the rump, and after a few seconds it falls over like a dog, twitching one of its back legs in gruesome delight. We all laugh heartily at all the attention the pig gets, lying through our teeth politely in that wonderfully British way. My back hurts already. My daughter is weighing me down. I suspect the pig knows this and is waiting.

Steve leads us on a merry tour of his farm. He shows us the chickens (Olly chases them all away and ravishes himself on the seeds before tipping over a bin full of animal seed, on which he then gorges). Steve shows us the horses (Olly nudges Steve so hard in the hip that he drops the horses' carrots, which Olly then dutifully snaffles). Then we see a newborn calf (Olly manages to knock over an electric fence, which takes 10 minutes to repair). 'Christ, Olly . . .' says Steve, rolling his eyes. After 30 minutes, each of us has become thoroughly sick of Olly. Even Billie, our faithful border collie, avoids eye contact with him.

We walk back along the fields, towards the pig-pen. A tiny voice upon my shoulders gives me real-time progress on Olly's whereabouts. 'Daddy, Olly's behind Mummy.'

'Daddy, Olly's stopped.' 'Daddy, Olly's chasing us again.' 'Daddy, Olly's going to get me.' My shoulders are killing me by now.

We stand at the edge of the pig-pen, and at this point Steve tells us a heartfelt story. Apparently Olly's mum accidentally sat on her offspring, suffocating all of them except Olly. They got Olly out, barely alive, and chose to nurse him themselves as a piglet in a shoebox, through to a small pig in a cage, then to this sub-adult, the whirling dervish around our feet. We almost feel sorry for him. It's the closest all of us have ever come to feeling sad about this immense barrel of a beast's sad little life.

And then. Olly manages to butt his log-like head through the fencing of the pig-pen and runs in; almost two-legged, he throws himself onto one of his cousins. His chubby mouth can barely open, so monstrous and swollen is his face, but somehow, by peeling his snout back over its head, he has the trigonometry to snap his jaws down onto the throat of this other pig. It yelps in a mix of terror and ill-fortune; even Olly's cousin hadn't seen this coming. We can't see the blood, though there is foam and saliva. Plenty of it. Olly throws his cousin left and right, then pulls him to the floor while second-by-second readjusting the location of his death-bite. We stand there in awkward (perhaps *the* most awkward) silence while Steve throws himself over the fence, pulls off a welly and starts smacking Olly over the head with it. 'DROP! OLLY DROP HIM! OLLY!' Round and round they go, Olly doing what the hell he wants, while Steve whacks him with his welly over and over again. It is horrible. Awful, really. All of us watching are dying small deaths and, though I want to help, I feel I can't with a little girl sitting on my shoulders.

Olly eventually loses interest and lets go, while Billie the collie manages somehow to shepherd the beleaguered cousin-pig to safety. Olly continues to take the welly to his head for a few more moments. Then the chaos reaches fever pitch. Suddenly Billie the collie, unable to control her excitement at all this – a situation for which I imagine she has never been trained – starts sprinting around and around and in between us. She looks the whole time at Olly. She starts panting and yapping, nipping at the heels of Olly who suddenly seems switched on to this activity, perhaps the only true danger he's ever been close to. Billie pushes, nudges, whines and spins Olly around. 'BILLIE. NO! BILLIE. HEEL! BILLIE!' And then . . . it happens. Billie inexplicably chooses to mount Olly. She pops her front legs over Olly's body as he lollops along, manages to secure some sort of grip with her paws, and then humps him for dear life. Olly manages to make it 30 or 40 metres across the field with this humping maniac on his back. Billie begins to pant more fiercely. It's almost as if she has an erection . . . Billie has an erection. Billie . . . is a boy. 'BILLIE, NO!' screams Steve. An angelic voice comes from my shoulders. It whispers in my ear quietly: 'Billie's being a naughty boy, Daddy.'

*

It is easy to drive these lizards down to any little point overhanging the sea, where they will sooner allow a person to catch hold of their tail than jump into the water . . .

writes Darwin in his *Voyage of the Beagle*.

I carried one to a deep pool left by the retiring tide, and threw it in several times as far as I was able . . . I several times caught this same lizard, by driving it to a point, and though possessed of such perfect powers of diving and swimming, nothing would induce it to enter the water; and as often as I threw it in, it returned in the manner above described.

We often think of Darwin as a rigorous scientist, a deviser of methods to determine truth, and a deep and careful theoriser and thinker, but I rather like this passage because it nods at his youthful curiosity and inquisitive vigour. What's most wonderful, of course, is that here we are, 180 years after he wrote those words, living in an era of Darwinian enlightenment that he, at the time, could never have foreseen and was yet to understand. A world in which we now apply Darwinian thinking to iguana hearts, iguana brains, iguana behaviours, iguana cells and iguana cell processes. And, of course, iguana masturbation. For this has become one the most famous of animal masturbators.

The marine iguana is a stocky animal, unique among modern-day lizards for living and foraging mainly in the sea. Darwin's description of them was surprisingly downbeat: 'The black Lava rocks on the beach are frequented by large (2–3 ft), disgusting clumsy Lizards. They are as black as the porous rocks over which they crawl & seek their prey from the Sea. I call them "imps of darkness".' That's how he put it in his *Beagle* Diary, anyway. In reality they are a little more colourful than this, acquiring tinges of red and teal, or brick red, depending on which island subspecies you observe. During breeding, males form little harems of females, which they defend diligently (males, as a result, have evolved a longer and more bulky size). Theirs is an arena not dissimilar to those of male elephant seals on their breeding beaches. And like the elephant seals, this puts the smaller, less dominant males without harems at a disadvantage. For them, sex on the sly may be the only hope they have of realising their genetic dreams.

In a 1996 paper, 'Pre-copulatory ejaculation solves time constraints during copulation in marine iguanas', the iguanas were to have their day in the sun, thanks to the report's authors, Martin Wikelski and Silke Baurle. The big problem for smaller male marine iguanas is interruption. Copulating toward ejaculation might take all of three minutes, which is a long time on that exposed, barren rock. Almost 30 per cent of males never make it, sent on their way by rival males angry at such ill-mannered interlopers. What the authors reported, though, was surprising. Some of the smaller lizards were observed to masturbate before encounters with females, readying themselves for sex – by doing this they could afford shorter copulations, less likely to be interrupted by rivals. According to the authors, if you're a young iguana, such pre-coital manual arousal is likely to increase your chances of a successful mating by an impressive 41 per cent; easily enough to be evolutionarily significant as a strategy. 'This tactic demonstrates the adaptive significance of a trait that is functionally equivalent to non-ejaculatory 'masturbation', and appears to be unique for vertebrates,' say the authors. A win for wankers, in simple terms.

Now, I know you're thinking it, and it's something I wondered, too: exactly *how* do marine iguanas *actually* masturbate? According to some personal correspondence with the author, 'The (mostly young male) iguanas assume a copulatory position – sideways bent back, cloaca pressed against a rock (usually). They do not make many movements, but some tail/hip pumping.' So there you go.

Iguanas aside, masturbation remains a bit of a mystery in Darwinian terms, for almost everything else in the animal kingdom. After all, think about it. In males, how could adaptations for wasting sperm (as masturbation sometimes entails) spread through a gene pool? And for females, what could such behaviour add to their reproductive success? Bluntly, how could rolling around pleasuring yourself lead to more offspring for those that do as opposed to those that don't? In fact, there are bigger questions to ask: is masturbation pleasurable for non-human animals at all, for instance? Is masturbation 'fun' for them?

Though Darwinian explanations, iguanas aside, do appear tough to come by, masturbation (often termed 'autoeroticism' in the literature) is widespread and commonplace, among vertebrates at least. It is known in a host of animals, both male and female. Lions, primates, bats, walruses, mule deer, zebras, mountain sheep, warthogs, hyenas, whales, birds. Animals use their flippers, their tails, their feet, their mouths for such purposes. Loosely, if it's reachable, then great: do it. They rub their nipples (if they have them), grab at one another's genitals, or rub their bits on inanimate objects like twigs (porcupines) or rocks (elephants and penguins), as well as animate objects like scuba divers (dolphins), for that matter. At the exact moment I write these words there is a YouTube video doing the rounds that shows a dolphin pleasuring itself into a fish's head. I kid you not.

My own observations of animal masturbation come mainly from our childhood dog, Biff. Biff was a prolific masturbator and used a number of tools to bring himself off: tennis balls, his mouth, cushions, rugs, his back leg. That sort of thing. You no doubt know dogs like this, too. But there was more. Though Biff occupies many of my memories when thinking about animal masturbation, I also remember a hippo I once saw in a zoo when I was little. While the young zoo-keeper was giving his talk to the watching crowd, the male hippo proceeded to stand behind him and have sex with a split watermelon. It was a strange scene that, perhaps unsurprisingly, has etched itself into my childhood. At the end of his talk, when the keeper asked the crowd, 'Does anyone have any questions?', about 30 hands went up. 'Not about the melon, please,' he added quietly.

Being observed often in zoo animals, such behaviour was widely assumed to be a clinical problem, like a nervous tick that blossoms in bored or abused animals in captivity. And it might be that, for some individuals, this is exactly the case. Brian Switek's memorable 2013 piece for *Slate* magazine entitled 'Sea Otters are jerks. So are Dolphins, Penguins, and other adorable animals' tells the story of a series of suspicious deaths of harbour seal pups in California's Monterey Bay. In the space of only three years, numerous cases were reported of male sea otters pleasuring themselves sexually with these pups, often injuring them fatally in the process. Veterinary analysis showed injury to the seal pups' flippers, eyes, noses, genitals and rectal tracts – all caused by over-exuberant sea otters. No one's sure quite what led the sea otters in question to behave in such a way. What stands out from the article, though (itself based on a technical report in *Aquatic Mammals*), is the identity of a pair of repeat offenders in these incidents: two rehabilitated sea otters from Monterey Bay Aquarium, once part of a programme to help get stranded and injured otters back into good health before being released. The suggestion is that young males, or those lacking an ability to compete, may search instead for 'female surrogates': objects on which they can fulfil a full range of sexual behaviours. Who knows, perhaps such behaviour could be considered a form of 'practise'? In which case there might be Darwinian benefit to such activities, as perverse and horrible as they seem to human sensibilities.

Though it seems to be fairly widespread among vertebrates, masturbation in animals is a difficult study area for those observing wild populations. It's not always easy to observe; it may be quick or fleeting, or hidden in other behaviours like grooming or scratching. As a result, such studies often bring up more questions than they answer. In September 2010, a paper in *PLOS ONE* reported on the wild antics of 20 male squirrels, all of which masturbated to ejaculation, with many then going on to consume their ejaculate afterwards. Quite simply, why? Why would they do this? What adaptive reason could there be for this behaviour? No one knows. Maybe squirrel semen has antibacterial properties? Or maybe drinking it helps prevent moisture loss? Maybe it cleans the pipes? Or maybe squirrel semen tastes nice if you're a squirrel? With a bit more time and effort these might become scientific questions or a call-to-arms for the next generation of scientists; questions against which hypotheses can be thrown, ideas tested and a truer vision of reality gained about masturbation. But it's early days.

At the moment, many observations of animal masturbation in the wild seem to be just that. Observations. But those marine iguanas stand out because they hint at an evolutionary purpose for masturbation. That's not to say, of course, that all masturbation is Darwinian, not directly anyway. Like any adaptation, sex can be commandeered and used for other purposes; namely, in one famous species, making friends of allies, and allies of enemies. And so, inevitably, it is time that we turn to bonobos.

The bonobo came to the world's attention as recently as 1929, when it was realised that the 'chimpanzee' actually represented two species. And it was only after the Second World War that their rich social lives began to be studied and understood. An early description from the time (based on zoo animals) outlining the differences between chimpanzees and bonobos holds as true today as it ever was. It noted that bonobos are lively, nervous and sensitive – lacking the fiery temperament of the chimpanzee – and that bonobos rarely fight; if they do, they kick rather than pull and bite. 'The bonobo is an extraordinarily sensitive, gentle creature, far removed from the demoniacal primitive force of the adult chimpanzee,' says one description. And, according to another early account, there was another observation: that bonobo copulation is 'more hominum' (yes, like us) and chimpanzees 'more canum' (that's 'doggy-style' to you and me). We weren't alone in being users of the missionary position, it seemed, which was actually rather a hammer-blow to some of those of a religious disposition. And those early authors, Eduard Tratz and Heinz Heck, noted something else: that the genitals of the female bonobo appear adapted toward this mode of copulation. The vulva sits between the legs (as in our own species), rather than protruding from the rump, so to speak (as in the chimpanzee). Here, among the bonobos, the clitoris is prominent, erectile and sits facing the front. Like us, then.

This was big news. After all, face-to-face copulation was

one of those things that humans were supposed to be alone in doing; a non-animal activity of great significance. This is, after all, the *missionary* position. This was sexual health advice spread via missionaries across the world to 'primitive' cultures. Only *we* could look into each other's eyes during sex, kissing, holding onto one another's bodies. We weren't animals! We were special! And so are bonobos. Bonobo sex is now well described, from both wild and captive populations. Even if you know all about it, it's nice to revel in the details every now and then, so please allow me that privilege for a few moments.

Bonobos are wildly promiscuous, both between and within sexes. Females may cup the bodies of other females in their arms, furiously rubbing their clitorises together (in an activity called GG-rubbing), males perform flagrant fellatio upon one another, or rub their penises together in a missionary position. Some have even been spotted 'penis-fencing' while hanging from trees – simply for the thrill, one would imagine. A classic is the occasional bit of 'rump-rump contact', where bonobos rub their behinds together to stimulate their genitals (rumpy-rumpy-pumpy, in other words). And they kiss, too. And I don't mean in a PG Tips way, I mean full-on French kissing. But that's not all. They are wonderfully open about masturbation, performed either on themselves or on others. They massage each other's genitals and they massage their own (though ejaculation from such endeavours has yet to be reported). According to observations, the most prolific masturbators are adolescent males and adult females, but, essentially, they're all at it.

If there's any sadness I feel, it's only that we discovered bonobos so late in the age of science. Imagine if we'd discovered bonobos *before* chimpanzees? How would our perceived ancestry look to us now? Would our history books feature pictures of ancient human ancestors less as primitive war-faring savages and more like a caravan of sex-hippies?

In fact, the 'make love, not war' slogan is particularly apt for bonobos, being as that's exactly what they use their sex for – peace.

In social animals, nothing stirs up aggression between individuals quite like food. Competition for food brings out the worst in them, and in most species you can expect squabbling to break out, or a bit of subordinate-slapping at the least. But in bonobos such scenes are rare. Why? You guessed it, they simply simmer the tension with a bit of GG-rubbing, mutual masturbation or full penetrative sex with their nearest and dearest. It's the animal equivalent of two British gentlemen trying to get into a lift together: 'No, you first.' 'No, YOU first.' 'I insist.' 'No, *I* insist . . .' Or that person in a business meeting who insists on everyone having coffee or tea the minute things get heated. Or the person who breaks up a fight in a busy bar by saying, 'come on lads, we're all here to have a good time.' Bonobo sex is a bit like resolving each of those scenarios – except with a spot of mutual masturbation.

Bonobos use these behaviours as a social lubricant (so to speak), easing the tension, promoting sharing, keeping everyone happy. But sometimes the females use sex as a weapon, of sorts. They offer up a bit of sex in return for something like a fruit that a male might be holding. It's what keeps the males in check, and makes their mixed travelling parties stick together – a situation, presumably, conducive to gene flow.

As I write these words, I have a weighty reference book on my desk, *Bonobo: The Forgotten Ape*, by primatologist Frans de Waal. It is a wonderful book, not least for Frans Lanting's pictures: page after page after page of bonobos looking playful, mournful, joyful, spirited, bouncy, appreciative, happy, wise, sensitive and occasionally, wonderfully, orgasmic – whether male or female. This is an ape that employs sex in the best way possible – a better life for all. A life where lonely prudes go extinct.

Not all masturbating animals should be thought of like marine iguanas, steely calculators of reproductive games of chance. If bonobos tell us anything about masturbation, it's that sex acts don't always have to be directly Darwinian – the reproductive benefits can be delayed. But masturbation might, in some animals at least, be less clinical; an emergent property of mismatched sex ratios, or of being a lower-quality male or female, itself perhaps a result of previous injury or incomplete rehabilitation (as may have been the case for those sea otters). Masturbation is complicated.

So what of Billie, the collie with which I began this chapter? The collie who suddenly, and without warning, took the opportunity to leap aboard Olly the pig and give him a damn good humping. What, exactly, may have been going on there? Why, exactly, do dogs hump? I've spent rather a lot of time pondering this question, and I haven't got many answers.

Dog experts tend to disagree about how influential humans were in domesticating our canine chums, but we do know through archaeological excavations that it probably took place somewhere between 10,000 and 20,000 years ago. How exactly they scavenged their way into our hearts, or to what degree we bred them for our needs (security, hunting, companionship) remain topics of serious debate. It all started with a wolf, though – we can agree on that.

Grey wolf sex is remarkably touching and has been the subject of a host of nature shows, some of which you're likely to have seen. Generally, grey wolves are monogamous, often pairing for life. In late winter the females become receptive; they start to carry their tail in a manner that shows off the vulva that bit better. During copulation, the male straddles the female from behind and inserts his penis, which he inflates a little more inside her. At this point he begins humping. They form a 'copulatory tie' – a linkage so tight that the male can, if he wants (or if he's pushed by a rival), flip round and stand back-to-back, with his penis still firmly plugged in between her legs.

If you were to compare and contrast the sex lives of wolves and 'village dogs' (a model African dog 'breed', considered most similar to that putative early dog-ancestor), it would be interesting to note a couple of adaptive changes. Village dogs apparently 'lock in' tighter when copulating, and may be harder for competitors to wrench free. They are also more likely to desert a female and look for mating opportunities elsewhere after copulation, in a way that wolves generally don't. In village dogs, it's all a bit more fluid, more competitive and more screwball, basically.

Humping is not restricted to males or females, and first appears early in many a dog's life, often through play. Experts associate it with excitement, arousal or stress and anxiety, with some referring to it as a classic displacement behaviour (an emergent behavioural response when a dog has conflicts of emotion). It might be triggered by a new and unfamiliar visitor, children, fireworks, house moves. That kind of thing. In these situations they're soon humping your leg, or in the corner humping a stuffed toy. In some dogs – according to the numerous web forums on the subject at least – humping may be an activity associated with relaxation; part of the wind-down routine, so to speak. But humping is also known from 'wild' observations of domestic dogs that have gone feral, exhibiting sexual behaviours.

In classic studies of the breeding behaviour of New Jersey's free-ranging 'wild' dogs the presence of an oestrous female led to increased aggression and a formation of dominance ranking among males that congregated nearby. Crucially, familiarity appeared to be an important predictor of mating success, and newcomers were more quickly ousted than familiar dogs. Studies like these suggest that dog days are rarely lazy; instead, each individual is under constant pressure to assess dominance and maintain contacts with those with whom they regularly meet. Loners don't appear to prosper, it seems.

All very interesting, you may say, but . . . what about the humping?

Popular parlance (and pet forums) have it that dog humping behaviour is about dominance, but quite frankly, I've failed to find out where this suggestion stems from. Is it one of those facts that is too good to check? If so, dog scientists really ought to talk to cat scientists. For in cats at least, humping behaviour is a little clearer and better understood, thanks to some sterling research undertaken in 1993 by Akihiro Yamane on a population of feral cats living on Aino-shima Island, off Fukuoka in Japan. Yamane's aim was to determine whether humping – of the male-on-male variety – was associated with one of the following: cluelessness, play, dominance, the inhibition of conspecifics (think: cock-blocking), or downright frustration at failing to get laid.

Of the 74 sexually mature cats in his study population, each had a name, so the sexual behaviour of each male and female could be monitored. How much humping would there be? Why did such mountings occur? The results gave more than a few hints. The most important was this: in 1,420 hours of watching his feral felines, Yamane never once saw male cats exhibit same-sex humping behaviour outside the breeding season. So in cats, humping wasn't about play. It was about sex.

But that wasn't all. He deduced that male-male mounting couldn't be about dominance either, since dominance is a year-round concern for cats, and same-sex mounting wasn't observed all year-round.

Another curious finding was that the 26 (yes, not many) cases of male-male humping behaviour he observed all took place in the presence of a nearby oestrous female. This seemed important. Female cats appeared to be part of the story; they had something to do with why some of these male cats were humping other males (on more than one occasion – and I love this observation – dominant males fell asleep after following a female around for hours, woke up and noticed the female escaping, and decided to mount a nearby male instead).

Yamane's working hypothesis was that frustration at failing to copulate with a female was what caused male-male mounting behaviour to occur. As far as I know, this remains generally accepted as a driver behind at least some homosexual behaviour in male cats.

And so, we return to those humping dogs. I was surprised at the lack of information on what role humping might have in domestic dogs, and how frequent it might be. Is there any science out there? If there is I failed to find it. Perhaps it's too difficult to unpick the many behavioural adaptations at work behind dog behaviour? After all, many dog experts still debate whether domestic 'wild' dogs are capable of living in stable groups, or whether the process of domestication has somehow made them incapable of living together (making them 'asocial'). Broken, you could say. Perhaps humping is simply a sign of a broken home, as the Victorians believed when they saw animals masturbating in captivity?

As part of my research into dog-sex behaviour, and to give myself a little inspiration, I visited a dog show – my first ever. Nowadays they call this sort of event a 'Fun Dog Show' in Britain to try to get around the perceived pomp and stuffiness of dog-breeding circles. And in some ways, yes, it was fun. A well-groomed rabble, if ever I saw one. Pugs, Pomeranians, Papillons, Poodles, Pinschers. Standing there, watching the trophies dished out to those pampered pooches, I saw a momentary fleeting glimpse of evolution's possibilities, a whirring nebulous mass of behavioural fluff whipped into a messy froth by the possibly misguided (but well-intentioned) aims of human dog-breeders. In fewer than 500 years, we've gone to town on the basic village dog design. Its 'natural' behaviours may now be almost unrecognisable to us. But we may get glimpses as they furiously mount our sofas, our tennis balls, our shoes, or each other (or, indeed, pigs). Are these snapshots of ancient evolutionary behaviours, or glimpses of the genetic goo that comes from living, and being selectively bred, alongside humankind? Could dogs be like those rehab sea otters, or are they more like bonobos, using sex for social gain? No one is sure. Young scientists take note: your everlasting contribution to the human pursuit of knowledge could lie just over this monumental and mystifying hump. I have stumbled upon it, and failed to scale it. I bid you better luck.

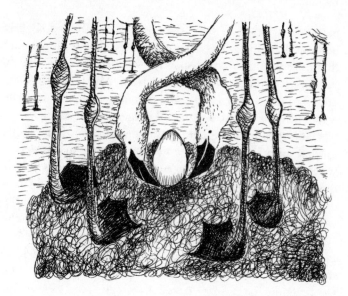

CHAPTER TEN

The Pink in Evolution's Rainbow

It's their legs I find most distasteful. I'm comfortable with the normal action of knees; of calves and ankles that swing backward underneath them but never forward. You can't bend your feet up and over your knees if you're a mammal; at least, I don't think you can. It's like a holy law. We just don't bend that way. But flamingos, they *mock* that law. They, and the wildfowl with which they sometimes cohabit, hock and spit at the rules. When I see them up close, it's all that I can think of. I'm trying now, really I am. In front of me stand 150 of the things, and all I can think about are their damned knees – or rather, their ankles (for that's probably a better description of what they are). Knobbly and slightly bulbous, like gristly bits of bamboo, with long trailing bones that look as if they could snap if I sneezed in a particularly strenuous manner in their direction. My binoculars are drawn to them, as they raise and then plop each foot into the water while prancing around the large enclosure.

Flamingos are camp. No doubt about it. In my eye-line, some stand motionless, arching their heads like anglepoise lamps in front of them. Every now and then they take a momentary pause: 'VOGUE'. Others move to and fro – crowing to one another, screaming, squawking loudly without apparent aim. A handful walk around on the currently empty nesting station to my right, an artificial island covered in specially made nest craters, which have since been improved by the flamingo residents, hopefully for breeding next year. A tall flamingo, like a pompous fashion critic, walks right up to the hide and flicks its head sharply left, to give us a good eyeballing. It's judging us. Me, with my unironed shirt and flappy baseball shoes and crap hair. I disgust it. It spins theatrically away from us, then walks a few metres before tipping its body forward and spilling its head into the water, filling up its beak with nutritious water-based spoils. After a moment or two it walks quickly away, as elegantly as a supermodel might . . . if you swapped her knees for ankles.

'We call them bomb-proof,' says my flamingo-expert guide. 'They're reliable. They breed well and so there's a massive age range in our enclosure; birds from every year for the last 20 years.' A voice speaks quietly to my right: 'And somewhere in here are Carlos and Fernando.' Carlos and Fernando: a world-famous pair of homosexual flamingos. This is a pilgrimage like no other.

It's August. I'm visiting the Wildfowl and Wetlands Trust (WWT) headquarters in Slimbridge, Gloucestershire. It's a wildlife park and conservation springboard – a site with collections of rare and exotic birds that incorporates an

internationally important wetland for breeding and wintering wildfowl. A kind of two-for-one conservation station.

Rebecca Lee, the flamingo expert, has kindly invited me
over to talk more about Carlos and Fernando, still arguably
the reserve's most illustrious pair. For someone who's
highbrow (she's Chair of the International Flamingo
Specialist Group), she's immediately amiable, likeable and
down-to-earth. With us is Mark Simpson, the WWT's
communications wiz. Mark is tall, handsome and rather
suave but, refreshingly, he's totally devoid of City Boy PR-
chat and happy instead to talk passionately about whistling-
ducks and celebrity flamingos. He's right at home here. We
make an odd trio, standing in the large, wooden flamingo
hide while thousands of young children and their families
spiral around us.

I cast my eyes over the crowd of greater flamingos outside,
trying to spot which, of the hundred or so, might be the
homosexual ones. It's tough. In 2007, Carlos and Fernando
made international news when it was announced that they
were successfully rearing a chick together as a homosexual
couple. A host of newspapers lapped it up, as did national
and international TV, including CNN (an experience
Rebecca describes as 'terrible'), plus a host of follow-up
articles and commentary about homosexuality in animals.
Even as I write this, Carlos and Fernando have just appeared
on the popular Facebook page '*I F*cking Love Science*',
attracting 77,000 'likes' in a flash. In all the press shots and
images they look every bit the celebrity couple, even though
they look EXACTLY like every flamingo you ever saw in
your life. I'm keen to hear about how they're doing now, six
years on from their burst into the public eye.

I'm enormously grateful that Mark and Rebecca have
taken the time to show me round. We talk a little about
flamingo breeding and the complications that it throws up,
before homing in on Carlos and Fernando and what, exactly,
happened here all those years ago. The first thing I learn is

that life is hard for flamingos. Before visiting I had assumed that they were gregarious, happy, sociable birds, but it turns out that they are anything but. Flamingo lives are fraught with neighbourly gripes. They obsess about location – namely the position of their nest within the colony. To be near the middle you need an iron beak: only the most aggressive, sizeable and single-minded of flamingos can manage the stress of it all. Most end up near the edges but flirt with death by predation as a result. THE EDGE is not a place that flamingos generally like, or aspire to be in.

Few of them appear ever to be happy with their lot in life, or their location within the colony, so they are understandably narky beings. They fight, squabble and challenge one another constantly during the breeding season, trying to obtain and secure the best spot, or at least a good spot in which to build a cupped mud-crater in which they may lay their single egg, their annual offspring. Though it may not be done on purpose, eggs are often smashed, or roll out of nests during the to-ing and fro-ing of flamingos coming and going from the colony. Victims of the din. So fraught is it that here at Slimbridge, Rebecca and her team often undertake something called 'egg-manipulation' – a technique to ensure maximum survival of the colony's offspring. 'If we've got a good dominant aggressive pair that lays an infertile egg, we might take the fertile egg from a pair that have since split up, and swap them,' she outlines. 'We give the good eggs back to the best parents.' This way, flamingo collections like those at Slimbridge can enhance the survival of chicks, and the reproductive success of the colony as a whole.

In 2006, Carlos and Fernando were just this: a good dominant aggressive pair with prime real estate in the middle of the colony. Excellent nest-builders, aggressive and fearless nest-guarders; that year, they got their surrogate egg – a move that paid off with a healthy, happy young flamingo. Their success was noted down in the records. Yet no one at this point had noticed that they were both male. It was only in 2007, after their selection as foster parents once more, that their 'history' was picked up on and the media cogs began to turn.

Though Mark wasn't at WWT at the time, he speaks with reverence about his predecessor's PR masterpiece, turning a simple same-sex pairing into a news tornado. 'It was a *huge* story,' he remembers. '*Enormous.*' He chuckles a little. 'Every now and again we get big stories, and generally these tend to be about breeding, but this was *big* – a global story.' He chuckles more heartily this time. 'I still find it amazing that just putting Carlos and Fernando into a search engine brings up reams and reams of stories about two homosexual flamingos.' There is nothing more annoying than coming into a PR job, with the big story of a predecessor still echoing around the newsrooms and blogs (believe me, I know). But Mark seems OK with it, genuinely revelling in the success of a press release that was, to all intents and purposes, remarkably simple: 'homosexual pair rear surrogate chick.'

It's a zoo press release that is actually rather more common than you might imagine. In 2004, Roy and Silo, chinstrap penguins, were announced as successful foster parents at Central Park Zoo, New York, to a global fanfare, and similar such pairing behaviour (albeit without offspring-rearing) has been reported from zoos in Japan and Germany. Indeed, researchers from Tokyo reported 20 homosexual pairs in a recent survey of Japan's 16 zoos and aquaria. So common may the behaviour be that it has even impacted on captive-breeding attempts for threatened species. At Bremerhaven Zoo in Germany, captive-breeding efforts for the threatened Humboldt penguins were hampered when their all-male pair refused to pair up with imported females and get on with the important business of reproducing for the sake of the species. German gay rights groups protested at the zoo's attempts to separate the homosexual pair (to which the zoo responded with sadness that 'nobody here wants to forcibly separate homosexual couples').

Buddy and Pedro – two male African penguins at Toronto Zoo – are another famous same-sex pair. Vultures, too, are known for exhibiting homosexual behaviour in captivity, the most famous being Dashik and Yehuda at the (ahem) Jerusalem Biblical Zoo. In 1998 they built a nest together and engaged in 'open and energetic sex'. As with Carlos and Fernando the zoo-keepers gave them a dummy egg to sit on, and then, after 45 days, replaced this egg with a hatchling baby vulture, which the two then raised successfully.

The media names given to these homosexual pairs interest me a little. Buddy? Pedro? Fernando? Carlos? 'So, who gets naming rights when a zoo notices homosexual animals?' I ask Mark, the PR guru. I don't know much about greater flamingos but I am vaguely aware that their distribution isn't as South American as the names 'Carlos' and 'Fernando' might suggest. Mark smiles. 'Yes, my predecessor in this post was a journalist.' He laughs. 'It might have been a case of – oh, yes, this'll fit, a couple of flamboyant, pink-sounding names for a pair of same-sex flamingos.'

We look again across the flock. Still neither Carlos or Fernando make themselves known to us. For a few moments the hide is suddenly empty of children and we talk a little more loudly, competing with the ringing barks and grating harrumphs of the flamingos that fill the window in front of us. If you pick out one flamingo at random, it's interesting to watch how it interacts with the others. I home in on a particularly small one, about the size of a little egret, and watch how it roams amid the legs of the older birds, as if navigating the roots of a mangrove forest. It avoids the loud or the flappy ones where it can, cutting a path through the flock that is anything but a straight line. Is it female? A male? Gay? Straight? Homosexual? Heterosexual? I admit to Rebecca and Mark that I'm struggling with the correct terminology for how, exactly, Fernando and Carlos should be best described. To me at least, it seems that academia has trouble defining such behaviours in human words – I'd seen a fleet of descriptives used in the media and in the academic literature before the visit. Is it acceptable to use such human words as 'gay' to describe animals? After all, aren't we all animals? And what are 'gay' or 'homosexual' animal activities anyway? Flamingo males rarely indulge in genital contact (females apparently occasionally do, though), so couldn't one argue that Fernando and Carlos are just playing house? And is it still homosexuality if an animal takes part in homosexual behaviours more for social reasons than for sexual ones (to garner an alliance within the group, for example)? Honestly, I'm not sure there's uniform agreement on such things yet. But in this chapter I'm going to stick with the usage of 'homosexual' to describe any same-sex behaviour – behaviour involving anything to do with courtship, copulatory behaviour, offspring rearing or simple touchy-feely between members of the same sex, particularly if the thing being felt is a penis or a vagina or another part of reproductive anatomy. You, however, are free to call it what you like.

I guess the lack of general academic agreement tells you a little about the infancy of homosexuality studies in animals. Things are changing, though. In zoological circles, studies of animal 'sexuality' are finally being given academic space, though slowly and with the occasional comic stutter. Where once upon a time homosexual acts in animals were considered an emergent property of some individuals within captive populations (as was also thought to be the case with masturbation), academic interest in homosexuality rose sharply in the new millennium, courtesy of a host of research articles and a few academic books that spelled out the depth and commonality of such behaviours in the animal kingdom.

We've already referred in this book to bonobos, and the females' fondness for GG-rubbing. And we've talked about why dogs hump, and looked at those penis-popping mallards and the Adélie penguins, both of which are noted for their same-sex liaisons. But look a little wider in the animal kingdom and there is more. Much more. Though accounts of animals totally eschewing reproductive sex with the opposite sex are rare (rams are perhaps the most famous example of this), observations of animals partaking in homosexual activity are common. Incredibly so. Bruce Bagemihl's 1999 book *Biological Exuberance: Animal Homosexuality and Natural Diversity* is the go-to text for anyone interested in such things. On his list are hyenas, lions, whiptail lizards, dragonflies, fruit flies, bedbugs, orcas, marmosets, brown bears, rats, cats, koalas, raccoons, barn owls, king penguins, mallards, ravens, seagulls, sticklebacks, graylings, char, whitefish, sunfish, leaffish, goats, salmon, garter snakes, geckos, skinks, rattlesnakes, desert toads and spotted salamanders, blister beetles, flower beetles, blowflies, houseflies, diggerflies and cabbage-white butterflies, octopuses, blood-flukes, chickens (and chicken fleas), creeping water bugs and . . . well, you get the idea.

Though accounts of such behaviour span the length and breadth of life's tree, it is among the mammals, and particularly among our own clutch, the primates, that homosexuality could be considered especially common. A quick scan through Sommer and Vasey's academic tome *Homosexual Behaviour in Animals* produces a sight to behold. The images that fill its pages are breathtaking. An American bison pounds its muscular hips into the butt-cheeks of another (for they, like European polecats, partake in full, penetrative anal sex with others of the same sex); eight fallow deer, males and females, frolic and thrust up against one another, huddled in the corner of an enclosure; red deer stags are pictured mid-combat, sporting stonking erections; CCTV images show male feral cats at it like, well, feral cats. Many of the images have arrows and circles added to them. Like something out of *Heat* magazine these arrows point out oestrous females, and pairing males ignoring them totally in the background. There are pictures of gorillas snapped in homoerotic encounters, sub-adult males staring at one another, touching and 'emitting copulatory whimpers'. And there, in the final chapters, our friends the bonobos appear. I am momentarily warmed by the sight of one male going at another, missionary-style. They're so refreshingly brazen. In fact the cover of this book is fantastic in its unsubtlety: it shows a female Japanese macaque tenderly riding the rump of another female, their eyes locked in love, lust or something we humans have yet to label.

Then there are the dolphins. Immature bottlenose dolphins are pictured writhing around in a big churning mass of white waves, one of them displaying a penis like a balloon-dog's tail ('The owner of the erection was not identifiable,' says the caption). Another shows an immature female with her beak up another female's jacksy (this is called, apparently, 'gentle goosing'). Bottlenoses are catholic in their tastes: males are known to attempt sex with turtles, eels, sharks and, predictably, divers. According to the book, bottlenose dolphins are as homosexual as they are straight: 'Few other species have homosexual interactions as often as heterosexual interactions,' say the authors. As well as gently goosing one another, their sex moves include 'push-ups', where one individual pushes up the genital area of another (often lifting them out of the water) and 'socio-sexual petting', when dolphins stroke one another's genital slits with their outstretched flippers (sometimes inserting them, too). To paraphrase the slogan of the UK gay-rights charity, Stonewall: 'Some animals are gay. Get over it.'

But *why*? Why do so many animals exhibit such homosexual behaviour? After all, there is something deeply un-Darwinian about homosexuality in animals. Males joining together, rearing the chicks of another female? Females forgoing offspring in search of some GG-rubbing? Goosing? Such things rub uncomfortably against Darwin's theory, which after all has reproduction at its heart. Why would nature select for an activity that leaves no offspring? How could the well-worn trope about the 'GAY GENE!' spread through a population without the mixing of sperm and eggs?

Now we're talking. This is the challenge that more and more scientists are taking on. Though debate rages, this is undoubtedly a fertile area of science and one in which a host of theories compete or work together. As with the bonobos and (perhaps) the humping dogs, one theory is that homosexuality can infer other advantages, like improving your social status or garnering alliances, activities that then increase your chances of 'successful' reproduction further down the line (such behaviours are considered 'socio-sexual'). Among the baboons of the savanna, for instance, males that showed off their genitals to one another, exposing themselves to the risk of having them scratched or bitten off, were found to form the best bonds and alliances. Ergo, they ended up 'fitter'. But there are other theories for such behaviour. Perhaps homosexuality can persist in a population if it confers a reproductive advantage on kin? Perhaps homosexual siblings help around the house, so to speak? Such an activity may, in theory, aid the proliferation of shared genes in nieces and nephews, even if they don't themselves carry on a direct line of descendants.

And what of humans, I hear you ask? I have no idea. For now, all I can do is repeat the fact that homosexuality is especially common in our primate lineage and leave it at that. Facts like this speak for themselves.

Reading the hypotheses that exist for the Darwinian benefits of non-human homosexuality, I can't help wondering whether or not, for some animals, it just feels . . . *nice*? Undoubtedly, though, in some species, it may be that some homosexual animals are simply making the best of a bad situation. One such creature may be the razorbill, a species of seabird, the females of which are observed to form pairs with one another only when males are scarce. Though not as successful (offspring-wise) as male-female pairs, it's better than doing nothing at all. If homosexuality does have a genetic component, it may be less of an evolutionary problem if animals can be sometimes homosexual and sometimes heterosexual – an activity that still allows such genes to persist.

The truth, I guess, is we're just beginning on this. We're only now at the point where we realise that some of the Darwinian jigsaw pieces may fit together. Now it's a case of trying to work out *how* they fit, a process that will take some time – and involve many more hours of fieldwork, awaiting those fleeting moments when a bison rears atop his conspecific or a bottlenose dolphin gives another a damn good goosing. We're on the way. Where once studying such behaviour made scientists fear for their careers, now, it seems to me, reputations are ripe for the making in the field of animal homosexuality – especially since it might tell us much more about our primate past (and primate present).

Later, as we sit in the WWT office with a cup of coffee, I ask Rebecca and Mark why they thought the flamingos made such headlines. 'It's our human obsession with sex,' answers Mark bluntly. 'And flamingos,' adds Rebecca, before reeling out some facts. 'Flamingos are in 70 per cent of all zoos, and they're one of the most popular birds. The public likes them,' she says. 'And they're pink?' I offer, before continuing, 'Are they the perfect storm in terms of stories of homosexual animals?' 'Yeah, there's a bit of that.'

I ponder this for a little while. I wonder at that moment whether flamingos are a kind of reflective prism through which we see ourselves, like gangly pink social commentators. 'Maybe,' Rebecca responds. Mark tells me of another popular story the public had recently devoured about the breeding habits of birds. Apparently the Bewick's swans that visit the reserve are renowned for their monogamy, with each paired up consistently year after year. Then, last year, the two members of a pair came back, both with new partners. It was the first time that such a thing had ever been recorded. 'We called it a "swan divorce" in our press release,' Mark smiles, to which I laugh. 'People like to anthropomorphise. They want to see a piece of us in the lives of animals. It grabs them.'

As we finish our cups of coffee, Rebecca offers me a summary of the media phenomenon that Carlos and Fernando momentarily became. 'I suppose what I find really strange is how much of a surprise it is to everybody that there are same-sex relationships in birds out there,' she says, looking out of the window at the crowds of people outside. She puts on a silly voice, '*Gay? What! Really?* YES, REALLY!' she shouts. 'They're all at it. There's loads of it going on out there.' Her tone changes and she becomes more professional for a second, sounding almost impatient. 'In the science or zoo community, flamingos do funny things; there might be two females one year, two males one female, trios, quads, whatever. Either way – *they all just get on with it.*'

If ever there is an opinion upon which science and culture may one day rest, it is with words like this. A non-controversy, no surprise, no front pages – nothing except a field rich with study, exactly like any other. An academic equality. A cultural equality. They all just get on with it.

I like the fact that during my visit no one could tell me which of those lanky pink birds were Carlos and Fernando, the homosexual megastars. I love that they all looked as racy as one another, hiding their celebrity status behind their strutting, nervous flapping and restless honking. I love that underneath their pink feathers, they're as interesting as anything else.

We'll know we're getting somewhere when, in future, homosexual flamingos will be called names like Stephen or Robert. Perhaps a measure of a future society should be what it calls its homosexual zoo animals? Maybe in future they will no longer need to be named? It's early days, but change is afoot.

Homosexual Behaviour in Animals (the book with the goosing dolphins) finishes with a rather lovely quote from Seneca, the Stoic philosopher from 2,000 years past: 'Nature does not bestow virtue; it is an art to become good.' These animals have a great deal to teach us yet about the human experience, and it's humbling to see that we're finally starting to listen.

CHAPTER ELEVEN

Mite of the Living Dead

Once, many years ago, I was giving a talk at a conference, proclaiming the importance of amphibians and how valuable they are for our gardens. 'BEST OF ALL, FROGS EAT SLUGS!' I proudly declared at the end. Bang. Up went an older gentleman's hand at the back. 'Excuse me, young man,' he barked. 'What, DARE I ASK, have you got against SLUGS?'

Slugs often suffer like that. Forgotten, overlooked, and too often treated like a second-rate denizen of planet Earth. It's hard to imagine us sharing a common ancestor (no matter how ancient) with something so alien to our way of life. Forget legs, they seem to say, why not surf on slime?

Forget teeth, why not invest in rasping? But it is their sex life that is arguably most alien to us.

Lacking bones of any sort, their sexual behaviour is less like watching machinery and more like observing amorphous slimy shapes interacting in new and novel ways, like blobs within a lava-lamp. The leopard slug (an easy one to find in most gardens) is a classic example, and it's one I'd like to tell you about, if I may. Where spiders leave their chemical 'come get me' clues in their silken webs, these slugs choose instead to use their slime trails. If another individual comes across this scent it's game on.

In the dead of night, the lowest of low-speed chases then ensues across your garden, until eventually one slug slides up behind the other, quietly nibbling the leader's behind. For their sex, they like a good overhanging object, often a branch or, in my backyard, a hanging basket. They start sliding up and down one another, twisting and spiralling acrobatically. Suddenly the pair of slugs slide down the branch (or whatever overhanging object they choose) on a trail of stringy slime like a bungee-cord, still twisting and entwined. They hang there for a bit. But things have only just started. Next, a big white male organ comes out from behind each slug's head, looking a lot like the snot that hangs from a toddler's nose after a surprise sneeze. Their genitalia then puff up, and slowly they get longer and longer until they match the length of the slug itself. Then (because, frankly, why not?) these appendages too start winding around each other, creating a kind of snotty helix that dangles from their entwined bodies.

Eventually the tips of these organs meet and clasp together, and each organ fans out to create a spiralling flower-like shape. Sperm is transmitted between them as the two leopard slugs hang there spinning, their bodies swinging and their appendages dangling from their heads, intertwined. They are both fertilised, for they (like most slugs and snails) are hermaphrodites; they come equipped with parts for giving and receiving sperm, and both sexual partners go on to lay eggs. They hang there for a bit longer, like some sort of sex piñata, then pull in their bits and shake each other free before falling to the floor and going their separate ways. Meanwhile we sleep upstairs, none the wiser.

We don't have much of an outdoor acreage, my wife and I. No big garden, just a courtyard, with the pond and some birdfeeders. What began as a stop-start spring has turned into a steaming summer, and we've taken to enjoying a quiet evening tipple under the stars while sitting on an uncomfortable white bench in the corner. We talk about our jobs, the sex research, our family, the stars – you know, normal stuff. Except on many occasions there is something else competing for our attention. By September it's hard for us to enjoy a quiet drink out there for all the noisy scratching of radulae on the fallen birdseed (listen carefully next time you're near a slug; really listen, and you can hear its scrapy 'tongue' scraping away as it feeds).

Anyway, hold on: there's a reason I'm telling you this. Whenever I hear this noise I normally have a look at what's going on, shining my little torch across the scene. SCRATCH. SCRATCH. SCRATCH. Sometimes there might be a tiny one. Sometimes there might be one enormous one, scraping itself up against a spent strawberry or a long-dead pistachio shell. SCRATCH. But last night I heard something else. It was a SCRATCH, SCRATCH . . . SCRATCH, SCRATCH. A pair of scratchings. Unusual, I thought. I shone my light underneath where I was sitting, and there it was. Or there they were. Two slugs, face to face, pumping slime out of their front ends in anticipation, I hoped, of some sex. I got my video camera out (oh come on, wouldn't you?) and focused right up close, right up and into their slime factory grille. On the viewfinder, it was like watching a time-lapse of candle wax pumping out of a sausage maker. Sickening and, sadly, completely non-sexual. Drat. No, they weren't having sex. Instead they were up to something else; they were midway through eating a third slug, one that I hadn't at first noticed. Two slugs mauling, frothing, radularising to a pulp a fallen comrade. Looking carefully, I could see the third slug's face, long dead but still moist, creeping out from between the goring mouths of the other two.

Later, when back inside with a glass of red, I watched my video on the TV. In many ways, it was disgusting. The video had a face-melting quality to it, not completely unlike those dying Nazis in *Raiders of the Lost Ark*. But hang on! Wait . . . what? WHAT? What was that? I rewound the video. Did I just see that right? PLAY. There! I rewound again. PLAY. There! On the screen I saw something impossibly wonderful. The two loping heads of the cannibal slugs filled the screen, their 'prey' between them. Slime pumped like a river of goo from out of their bodies somewhere off screen, as they rhythmically scratched and digested their slimy molluscan dinner. There was an audible scratching coming from one of them. Their feelers unravelled, tentatively touching one another . . . then . . . THERE! In the harsh light of the camera, and with the close focus, I saw something that took my breath away, perhaps more than anything in this entire year of sex-surveying. I saw an *animal*. It came out of a hole on one of the slugs' bodies and it ran around a little bit, like a tiny white clockwork car following an invisible circuit on the slug's back. And then it went right back into the slug's body, back into the same hole from which it first emerged (the pneumostome – the big one on the side).

I watched the rest of the video but failed to see this little robo-organism show itself again. I rewound the video all the way back to the start, one more time, to double-check I hadn't imagined it. Nope, it was definitely there. A tiny mite living on a slug.

I wandered back outside with my high-power torch and found the same slugs still feasting on the corpse. I shone my torch right in their faces once more, and waited. I witnessed that mite again. I watched it in real-time. And I saw others running around on the slugs' backs. Not one, not two, not three – 10, 12, 15 of these surprising little beasts. White dots that ran manically around on the slugs' bodies, popping in and out of the pneumostome, running up and down their flanks and along their backs. They even occasionally traversed the slug's antennae, like pirates clawing their way up to a crow's nest. And, perhaps most amazingly, I saw these little white dots jumping ship – running from one live slug to the other, using the dead slug in between like a kind of land-bridge. It was wonderful. I mean that: really wonderful.

There is a fantastic thing about being a naturalist. Apart from all the awe and wonder nonsense, there are those moments when you witness something you believe, wholly, that you may have been the only person in the world to notice. An albino tadpole. A dog performing auto-fellatio. Heron vomit. These things sound rare, but I was dealt crippling blows upon learning that each and every one is anything but. Not rare at all. But what of these little white clockwork vehicles coming in and out, and running all over, the slugs in my garden? What of them? Could I have discovered a new life form?

You know, with mites, it's not entirely out of the question.

Mites are a diverse bunch, fitting taxonomically within that enormous bracket, the arachnids. They are masters of the microhabitat. Specialisers beyond ordinary levels of speciality. Species of mite live only in the noses of seals, upon the legs of chickens, in the ears of porcupines, in the middle of a sea urchin or up a bat's bum. One flower-mite species even travels from petal to petal via the nostrils of a hummingbird. If you want to name a new species after a loved one, grab a microscope and a nearby animal, and plunge that microscope into any of the animal's orifices. If you can wade through the literature and convince the experts that it's new, then congratulations, you've found your very own mite.

In some ways, mites mirror the mammals in their diversity – there are climbers, biters, grazers, ones like all-terrain vehicles, swimmers, bloodsuckers, gliding ones, ballooning ones, and even ones that look a little like peacocks. The biggest, the Indian tick, is about the size of a ripe grape. The smallest is a fraction of the size of a full stop.

But what I like best about them is how spectacular their sex lives are. Of the 45,000 or so recorded species, nearly all those whose sex lives we know about are impressively gratuitous. Perhaps one of the most common mite behaviours is mate-guarding. Males often guard immature females and then, when the females moult into their adult coat, have sex with them. Some of the males have little suckers with which they carry the females around safely until they reach their sexual phase. In some species of mites the males may share little clutches of females if there aren't enough to go around. Harems, too, are common. If you've ever disturbed a freshwater mussel, it's possible you also disturbed the water-mite harem within.

Males fit lots of stereotypes in the mite world: there are the fighty ones, the dancy ones, the showy ones. Stephen Jay Gould tells a story (in *The Panda's Thumb*) of one kind of mite that lays a clutch of eggs, the first hatchling being the male and the rest – his sisters – the females. As the females hatch he wanders about inseminating each and every one of his siblings, before he promptly dies. Oh, and one more thing. This all happens inside the mother's body, which the now-impregnated baby daughters then devour from the inside out. The male has sex and dies before being born, essentially.

Even mite eggs, and the spermatophores that some species use to fertilise them, are diverse beyond belief. Each of the spermatophores is laden with adaptations to better guarantee successful passage into the female's body where (often) her unfertilised eggs sit. Stalked spermatophores, droplet-encrusted spermatophores, flask-shaped ones, blobby ones. They even indulge in diverse sexual positions – some mate face-to-face, some back-to-back, some at length, some like thieves in the night. If we ever discover life on another planet, my guess is that it'll be like this.

And of course, dear reader, those mites are quite likely to be there now, on your body. Having sex. On your face. Right now. Your face, to them, is simply an arena to their never-ending sexual dramas. To them, your face is beautiful. Even in the morning (for all we know, *especially* in the morning). These are the face mites. Their bodies are like long sausages, and when they're not looking out for sex, they like to park themselves in your pores. *Demodex*, they're called. Not all of us have them, but still . . . that first kiss your mum or dad gave you as a newborn? It was a potential land-bridge. The same goes for every sexual encounter you've ever had.

Don't worry, we're not the only ones. Sexual unions within a host of creatures provide useful land-bridges across which pioneering, sex-obsessed parasites can migrate. Whale lice are one such group. They are parasitic crustaceans (more properly called cyamids) that look a little like a woodlouse with the legs of a spider crab. Some cyamids are quite large; bumblebee size – certainly too big for our meagre bodies – and they scramble around on whales, often seeking shelter in calluses and genital slits, rewarded with their share of dead skin and algae. Some whales have as many as 7,500 whale lice living on them, so they can be pretty common, but the precariousness of their existence is underlined by this, one of my favourite-ever sentences in a conservation report: 'Sperm whale lice are not considered by the World Conservation Union (IUCN) to be threatened or endangered, although their only habitat, the sperm whale, is listed as Endangered, or facing very high risk of extinction in the wild'.

But I digress once more. This information is all well and good, but what about my little slug mites? Like many poorly trained naturalists, I reached not for an ID Guide at this point, nor for my library card, but instead went straight to Google. I typed in 'SLUG MITE'. Some 2,000 search results came up. The top one, of course, being Wikipedia.

> Riccardoella limacum *or the slug mite is a member of the Acari (mite) family which is parasitic on slugs and snails. Slug mites are very small (less than 0.5 mm in length), white, and can be seen to move very rapidly over the surface of their host, particularly under the shell rim and near the pulmonary aperture. While once thought to be benign mucophages, more recent studies have shown that they actually subsist on the host's blood and may bore into the host's body to feed.*

There's a spattering of information below this text, but no picture to speak of. Apparently the little things move from host to host mainly when the slugs mate, and even occasionally traverse the mucus trails they leave behind. Infestations impact on slugs, too: infected ones take longer to mature and appear less likely to mate and feed properly. But it's their sex lives that interest me, and on this subject Wikipedia was rather light on details:

> *Mites have two sexes. Their five-stage life cycle is as follows: Females lay eggs in the host lung, and then the eggs hatch in 8–12 days as six-legged larva in the lungs of hosts and undergo three nymph stages. The whole life cycle can take place in 20 days under ideal conditions.*

That, my friends, is it. Did they have sex face-to-face? Dunno. What's his spermatophore like? Dunno. Are the females guarded by the males? Dunno. Does the female guard the male? Dunno. Are pheromones involved? Dunno. Where on the slug do they have sex? Dunno.

No fear, I thought. I'll just Google something else. Something like 'slug mite sex'. With hindsight, the result is probably obvious: a big fat zero. Nobody in the history of the Google-scanned internet has put those three words together in a published web page. Maybe this is what it feels like to stand on the frontiers of human knowledge. It was exhilarating – I was a pioneer! I couldn't use the internet to answer a question. The internet was useless to me! Imagine!

I flipped the laptop screen shut, grabbed my phone and went on Twitter. It was time to find an expert. 'Anyone know a mite expert?' Silence. Zero responses. I tried again later. Zero responses. I spoke to specialists and their associations and societies. I spoke to entomologists. I spoke to arachnologists. I spoke to research agencies. Nowt. At one point I almost got a meeting with the global mite expert at the Natural History Museum, but it seems she had bigger fish to fry than my EARTH-SHATTERING mission to uncover the truth about how slug mites have sex. 'I'm not bad with my oribatids if you want to chat,' said one offer via email a few days later. 'I'm pretty good with soil mites,' said another.

I politely turned them down. I know I'd only be sitting there while we chatted, wondering not about soil mites or flower mites, but instead about my precious slug mites, diddling in and out of that slug's pneumostome, considering their secret sex lives. Thank God, then, that Britain's libraries have not yet been lost. I lapped up what information I could find and here, for you, is the regurgitated contents of that research.

Slug mites were first observed in 1710, and named in 1776 (in the same year as the dugong, the beluga whale, the meerkat, the hoatzin and the honey badger, if you're asking). It took almost another 200 years for them to be described more properly, though. It happened, just once and once only, in 1946, in the journal *Proceedings of the Zoological Society of London*. 'A Monograph of the Slug Mite – *Riccardoella limacum* (Schrank.)', by Frank A. Turk and Stella-Maris Phillips. I couldn't dig up much on Phillips but, according to Brazil's Museu Nacional, Frank Turk became interested in mites after a long period of ill health before the Second World War. A 'deep thinker' and immensely well read, he apparently also liked gardening, Siamese cats, music, art and poetry. Anyway, here's Turk and Phillips's official account of the slug mite, starting with the female:

> The skin appears perfectly smooth when examined with critical illumination under a good one-twelfth oil-immersion lens. The rostrum is short and broad, the chelae stumpy and somewhat skittle-shaped, and the three-segmented palpi have four short feathered hairs on the distal segment. The body hairs are short, thick and rod-like and are, moreover, very finely feathered for the distal two-thirds of their length.

It goes on.

> In the centre of the posterior half of the abdomen is placed the genital opening, closed by two shutter-like valves and bordered anteriorly by five very short spine-like, unfeathered hairs. Two pairs of genital suckers are placed laterally on the valves and outside of these are three pairs of somewhat longer spine-like hairs.

And the males?

> The most obvious difference between the two sexes is that the fully adult male possesses two additional pairs of genital suckers, much smaller than the other two pairs but quite distinct in most individuals . . . It seems therefore certain that this additional pair of genital suckers appears only after the last moult and in the truly adult condition.

There is something genuinely touching about reading such information for the first time, knowing that, on this day, I am possibly the only person on Earth thinking about slug-mite sex. I am surprised, though, at how little information there seems to be. Their sex lives still appear shady; hidden somehow. I scan through the paper.

It has not been possible to gather many definite facts about either the mode of coition or of fertilisation but this, of course, is a difficulty which confronts the worker on any of the prostigmatic terrestrial mites, for, in spite of the many observers, nothing is yet known about the method of copulation in the comparatively large, common and easily observed mites of the family Trombidiidae.

BLAST AND BLAST! My inner monologue curses. So near, so far. How do these little blighters have sex? ANYONE? I consider asking the librarian, but instead I take a breath and carry on reading. There, though, near the end, is a glistening ray of hope:

In June 1943 a male of this kind was observed on Limax maximus [a snail] *in what appeared to be* in copula *with a female . . .*

My eyes widen.

Observation was only possible through a hand lens as an attempt to lift the pair on to a microscope slide on the point of a fine brush was unsuccessful and they parted company. The female appeared to be raised a little on the two front legs so that the hind part of the body was higher than the fore part. The male approached (head facing in the direction of that of the female) and making a complete turn on the right-hand side of the female it took up a position, such that the tip of its abdomen was touching and slightly underneath that of the female; the fourth pair of legs touched and were possibly locked with those of the female whilst the third pair were out of sight underneath the body. They appeared to be perfectly stationary for about 30 seconds and then a slight movement – possibly a disengaging movement – was noticed on the part of the male.

Jackpot. Ish. The rest of the paper was made up with bits and bobs about the mite's anatomy and anecdotal stories of its behaviour (including the incredible news that they can get into breadbins via the snacking forays of nocturnal slugs, and that they sometimes form small aggregations, like flocks of starlings, all running around together on the poor, infected molluscs).

I searched and searched and found little else. 1970 was the last time the little sprites appear to have found a similar amount of attention, when they appeared in a paper by R.A. Baker in the *Journal of Natural History*. In this paper there is a host of diagrams of the mite's tiny genitalia, and its eggs and larval stages, but sadly it mentions nothing of the sex lives of the slug mites, only their anatomy.

> *The female opening is long and narrow, except when an egg is present in the posterior part of the female reproductive tract, whereas the male has an oval shaped aperture which is broader and shorter than that of the female. In addition the male has a more strongly sclerotized lining to this opening and possesses barbed setae in the genital vestibule.*

And that was the way it went. I had stood at the edge of human knowledge, asking sexual questions that could barely be answered. I had skimmed pebbles out to sea, and seen more ripples than I could count. It felt inspiring, actually.

I rewatched the video of my slug mite a few times. Everything you could ever want to know about sex you could learn from mites, so diverse are their lifestyles and so complex their sexual behaviours. They are nature's textbooks, viewable to anyone able, and with enough patience, to read the tiny font. Yet there is so much still to discover about life, and those slug mites prove it.

I once heard the great naturalist E. O. Wilson being asked by a journalist for any advice he'd offer to young scientists eager to pursue a career in the natural sciences. To paraphrase slightly, he told them of how, in the American Civil War, lost soldiers were told simply to head toward the sound of guns. His advice to young scientists was simple: 'Head *away* from the sound of guns,' he said. The words stuck with me. Find your own fields to plough, he was telling them. Make a name for yourself in something obscure (for him it was ants) and shake science from the ground upwards, as he did. Slug mites remind me of that. Slug mites: they are silent, unswept, virgin forests of biological exuberance. And they are out there right now, in your gardens, eating bits of the slugs that eat your strawberries, your birdseed or each other. With a microscope, a bit of free time, a bit of patience and an eye for the ghoulish anyone could be a global expert in a given species of mite. Sex on Earth is anyone's game.

CHAPTER TWELVE

The Greatest Story Never Told

At last he is alone with her, the object of his affection. He has followed her for more than a mile, glued to her trail. He has fought for her attention, vanquishing his foes with strength and valour and never once resorting to biting, eye-gouging or any such ungentlemanly behaviour, for he would never stoop so low. He is a male of honour. Integrity. He finds himself now in a secluded pocket in the grass, sitting beneath the trunk of an old oak. For the first time in their long journey through the woodland, she permits him to come closer. They touch for the first time. He is gentle; his movements kept slow for fear of losing her at this, the final hurdle. They take each other in, touching, smelling, almost tasting. He stops. Her chest heaves. He moves his head so close that her breath becomes his. He sweeps his long body over hers – the friction fizzes, his passion palpable. Suddenly they slow. As if in a trance, they roll and loop over one another's bodies. There is pre-coital frisson. Like her, he has waited years for this. These are snakes – adders – and this is their sex. Slow, considered, somehow valiant and, by chance, playing to our own literary interpretation of romance and steaminess, if only in a shaky 'Mills & Boon' way.

But, sadly, it's a sight few of us see in the UK. For the adders are mostly gone. Those that remain are often packs of sad loners, restricted to tiny pockets of heathland or ancient woodland. Isolated like this, they are like shipwrecked sailors on a chain of desert islands, drowning under the increasing water level of human interference. One by one, the islands are disappearing, and we should expect further losses.

Historically, the life story of our snakes is a suitably twisting tale: they are the surviving members of a lineage with its roots in the age of the dinosaurs, a modern fate chinked by Biblical betrayal, persecution on a national scale and finally, right at the end, some legal protection (in the UK, it's illegal to hurt or kill them, and about time too). Perhaps in the public conscience their reputation as an evil-fanged death-stick is slowly changing? Perhaps slowly people are coming to view such animals as beautiful? Indeed, it's fair to say that ours is the generation that is at least more likely to reach for a camera than a spade. But is it a case of too little too late?

It's at this point that I'll reveal a little something about myself. As is probably obvious by now, I'm a naturalist who is a tiny bit frustrated with certain animals. I'm full of love and passion for nature's unloved and forgotten, of course, but I'm occasionally resentful at the amount of time lavished on nature's popularists. You know the sort: Beatrix Potter's batch. Through her books, Potter created nature's first 'in-crowd' – a world where a bunch of charismatic characters have each other's backs; where the horseflies, the cockchafers and the water mites get stuffed. Squirrels, rabbits, badgers – they've become nature's jocks. They're where nature's at. And there, sitting resplendent among all of these popular creatures is my *bête noire*, the 'humble' hedgehog. The British public loves an accessible receptacle of grief and sorrow, an underdog. To nature fans, the hedgehog has become that. It's the Princess Diana of nature. In a recent *BBC Wildlife* poll to find Britain's first 'national species', the hedgehog walked it, with 42 per cent of the public vote. What can I say, we British love our hedgehogs. I sometimes wonder how much Beatrix Potter has had to do with this – whether her vision of Mrs Tiggy-Winkle stuck, somehow. That simple hedgehogian washer-woman, living in quaint Cumbria, delivering laundered goods to the woodland animals around her. Perhaps it's the social schtick that we're drawn to with such popular animals?

Not for the first time while researching this book, I'm drawn towards a subject area that may threaten to engulf me. But still . . . If we knew better the sex lives of these creatures, might it change our opinions of them? Are there myths to be busted about Mrs Tiggy-Winkle and her ilk? Could their sex lives be even more interesting than their public image suggests?

There seems to be only one person to ask. I collar him via email, and he invites me over to his house. I'm referring of course to a man, one they call 'Hedgehog Hugh'. If the hedgehog is the nation's darling, then Hugh Warwick is the nation's hedgehog darling. He's a true luvvie for hogs. Author of *A Prickly Affair*, in recent years he's become a mouthpiece for hedgehog-kind and a popular commentator on the nation's spiky sweetheart. When we meet he is largely as I imagined him – erudite, energetic and wonderfully passionate about everyday nature, especially his beloved hogs. He invites me through the door and into his world: a crowded living room, littered with hedgehog souvenirs, pictures, posters and bookshelf after bookshelf of interesting books, many of which, unsurprisingly, are about hedgehogs. Within minutes he has me standing outside in his garden, holding up a bit of cheese in my hand for a robin he has habituated to humankind. That's the sort of guy Hugh is. A friend to the animals. As he prepares the coffee, I stand there in silence with my hand open for this plucky robin. It doesn't seem to want to come near me. I stand for minutes with this little bit of cheese, wondering if this is some sort of bizarre initiation that I'm somehow failing. I wait, but no robins magically appear. I fail.

Hugh is unperturbed. 'No matter, he'll be back later,' he says, and we wander through his kitchen and back into his front room (I quietly put the cheese in my jacket pocket, fully aware that I will find it in three months' time and gag). We park ourselves down on the sofa and make small talk while I eye up some of his hedgehog paraphernalia – teddies, books, postcards and, in the middle, of course, a stuffed hedgehog. About the size of a slipper. I am momentarily struck by its ears, which look surprisingly like my own. Big, pink, shapely. Somehow . . . flappy. Do all hedgehogs have ears? I'd never noticed them before. They're almost primate-like; dare I say it – cute? Its spines are not fully erect, and I can see its four pokey legs coming out from underneath. Around its face and along its underside it has fur. It's furry. It *is* cute. But then . . . its beady, possum-like eyes momentarily lock onto mine.

I come clean to Hugh that I'm here with questionable intentions. Or rather, I fumble around a bit trying to communicate that I find hedgehogs . . . a bit, you know, annoying. I don't think he hears me. Instead he rummages among his many books and photo albums, readying himself to tell me all about his passionate encounters with hedgehogs. Encounters that span decades.

The first thing he tells me is that hedgehogs rut. They are rutters. This surprises me a little. Rutting is something I had assumed only big, powerful beasts could do, but apparently not. No, hedgehogs rut. In Britain, they begin rutting from around May onwards, sometimes earlier. Hugh pulls out an academic work called simply *Hedgehogs,* by Nigel Reeve. We flick through the pages on hedgehog reproduction. According to the book 'the basic sexual anatomy is unremarkable'. I jot this phrase down speedily. A diagram shows a side-on view of a male hedgehog's genital tract. There, among the glands, the vas deferens, the prostate, the bladder and the seminal vesicles is a long, stringy thingy labelled 'penis'. At the tip of the penis there appears to be something a little like a bulbous hairy scab or a cornflake, which is, bizarrely, the glans. 'It is surprising that more people have not remarked upon the observation by Poduschka (1969), that in young *E. europaeus* (about two months old), the penis tip is a dark olive-green, which changes to a more conventional pinkish-red in adulthood.' Surprising indeed.

I feel myself drawn to asking the inevitable, out of interest really. 'So, Hugh, how long is a hedgehog penis?' I used to blush asking such questions, but now, after many months of animal-sex exploration, I'm a hardier kind of pervert. It barely registers as weird any more. 'Hmm . . .' Hugh gives himself a moment, flips over the page and reads out a line from the book with a booming Shakespearian voice: 'The erect penis protrudes from its sheath by several centimetres but I have not found exact measurements in the literature. Hardly surprising in view of the problems presented in obtaining such data.' With that Hugh theatrically slams the book shut and puts it on my lap, while he goes to look for other information on hedgehog sex from more of his academic tomes.

I ask Hugh about female hedgehog anatomy, flicking further through the book he has placed on my lap. There, a few pages later, I come across a host of diagrams a little like those I remember from school: two ovaries, a cervix, a muscly vagina ending in a hairy vulva and clitoris too. The next page contains a handy flow-chart, which essentially tells me that, if the season's good and they aren't suckling, weaning, pregnant, pseudo-pregnant or midway through sex with another male, the females are in oestrous – ready to mate. 'And courtship?' I ask. Hugh smiles. This is clearly something he has much more direct experience of. 'Well, now this is where it gets interesting.'

From this point, there is a slightly rushed, excited quality to his sentences. 'Normally hedgehogs tend to avoid one another,' he starts. 'You don't have territories of hedgehogs, you have home *ranges* of hedgehogs, and they all tend to overlap quite considerably. BUT!' He draws breath. 'When a female comes into season, she will suddenly attract the attention of males, normally one but sometimes more than one. The males come along and the female stops what she's doing. She bristles up a little bit. She frowns.' 'Frowns?' I throw in quickly. 'She frowns,' he says with utter seriousness. I allow him to carry on.

Hugh uses his hands now, pretending one is a female hedgehog and the other a male. 'The male then tries to circle around to get behind her. He wants to have a sniff and to see about access.' One of his hands sidles up to the other nervously. 'If she's not interested in any of that, she will turn around to face the male – and she might periodically jump forward a little bit, and do a little sneezy noise.' Hugh looks up at me and bares his teeth, forcing the air up and out of his nose to make a little snort. 'TITH . . . TITH,' he wheezes. 'That's the noise they make, a kind of sneezy snorting sound.' It sounds authentic. He continues excitedly. 'And the male hedgehog will jump back at this noise. There will be a momentary calm, and then she might go back to how she was, but he'll keep at it, circling around until, again, he gets too close, and she'll turn around and do the same thing again.' He makes the cute sniffling 'TITH . . . TITH' noise once more, louder this time. 'They continue on like this, and it can go for hours and hours and hours and hours . . .'

I am immensely impressed at hearing this, having previously considered hedgehogs' sex lives to be somewhat less desperate and slightly more mundane. Being famously covered in spines makes hedgehog sex tricky. A happy female has her spines pointing backwards, slightly flattened. An unhappy female 'frowns', pulling her skin up and towards her head, erecting the spines, making it impossible for a male to get close to her. Unless he wishes to have his fragile genitalia painfully speared, he must ensure that the female is 'happy', or at least not grumpy – and certainly not frowning – when he makes his final approach.

All this circling around one another does strange things to the places in which hedgehogs dwell, apparently. Hugh rummages around in his files and pulls out a photo album that he tells me contains an image of something called a 'hedgehog carousel'. We browse through some of these photos, all of which are of Orkney, where Hugh spent a great deal of time studying hedgehogs in previous decades. He looks almost the same age in the photos as he does now, which unnerves me slightly, and in each picture he has the boyish grin of someone who can't believe that he's being paid to look at hedgehogs. There, near the end, is a photo of 'hedgehog carousel'. It's strange. A circular, flattened patch of tussocky grass, in which hedgehogs must have courted the night before. The constant attention from the male, aimed at the female hedgehog's nether regions, and her attempts to keep her eyes firmly on him make what looks like a perfect circle in the grass. There, in that photo, it looks almost as if a flying saucer had landed there – unexplained and somehow eerie. Hugh delightedly informs me that it led to the greatest *Guardian* headline ever: 'Hedgehogs cleared of corn-circle dementia.'

It strikes me as something worthy of literary interpretation, of inclusion in a novel for children AT THE VERY LEAST. I feel almost vindicated – my quest to dig up some dirt on Mrs Tiggy-Winkle is already bearing fruit, and I wonder for a few moments whether Beatrix Potter might have told a different story had she known all of this. A different story to that which she birthed into popular culture; that of the bumbling washer-woman turned crop-circle prankster. A mysterious, spiky myth-maker.

There are many other examples of creatures that have stepped from reality into popular culture, carrying with them sexual attributes given to them by human minds. One of the most beguiling, perhaps, comes from *Finding Nemo*, among the most popular children's films of recent times. In reality, the truth is far different from and, if you ask me, far more interesting than the fiction. In the movie's opening scenes we see a male and female clownfish tending to their crop of eggs, before she, the mother, is suddenly eaten by a barracuda. Nemo, the only surviving egg in the batch, is reared by his father before setting out on his wild adventure. But hold on. If the mother had *truly* succumbed to a barracuda, things would have turned out rather differently. The father would have changed, like many male reef fish, into a female. Sequential hermaphrodism. Being an only child, Nemo, born as an undifferentiated hermaphrodite, would have grown up to become a male and, in a neat twist, would quite possibly have had sex with his now-female father. But that's not all. Should the father later die, Nemo would continue the family trend by changing into a female and having sex with their offspring, should there be no other clownfish around. And there we have it. I have spent a lot of my life working with eight-year-old girls and boys. I think they'd love this story.

So what of other such movies? *Dumbo*? Imagine the scene where he turns into a green-penised monster during his hormone-powered 'musth' period of sexual activity, characterised by earth-shattering elephant-rage. *Lady and the Tramp*? It's still set in an alley but the cast is bigger and the film lasts a lot longer than 75 minutes. *The Wizard of Oz*? You don't want to hear about what the lion does when he wins back his courage. Hell, *Bambi*? Perhaps she was a non-reproductive male morph called a 'peruke', a gender known within a host of deer and elk species. I allow myself to imagine time-travelling, sitting down with the authors of these great books, telling them about what science now shows. *Winnie the Pooh*, *Flipper*, *Doctor Doolittle*, *The Jungle Book*: stories that would have ended very differently had they been based, even a fraction, on reality. Alas, I'm being facetious: they are just stories – I know that. And we can't go back. But I see no harm in telling the true stories, at least as a light centrepiece within this book. Allow me to return to those hedgehogs.

'Can we talk about vaginal plugs?' I ask Hugh. I'd read somewhere that male hedgehogs can leave a plug in the female's reproductive tract that stops the sperm of other males from sneaking in. In my travels I've heard it referred to a few times when talking about hedgehogs. Hugh sees where I'm going, but lets me down gently. 'Possibly, but then possibly not . . . really, it's a moot point.' He takes his time on the subject, relishing the mystery. 'Some people have gone through the make-up of the plug in great detail, and they've found that it's gelatinised ejaculate; others have gone through it and argued that it's as much to do with the female's epidemial cells as it is to do with the male.' Another mis-imagined meme, then? 'The truth is,' Hugh says with a smile, 'I really don't think many people are studying the intimate recesses of post-copulated hedgehogs.' Ahh, I say. Fair enough. As on so many other occasions during my year of research, I am suddenly aware of bumping up against the boundaries of current human knowledge.

Thankfully DNA tests on hedgehog offspring offer another way to understand their reproductive strategies. In studies of hedgehog genetics undertaken in 2009, multiple paternity was found in two of the five litters analysed. If this is true of all wild hedgehog populations, then sperm competition is likely to be at play somewhere within the deeper recesses of hedgehog reproductive anatomy. It's just a case of finding better ways of looking, perhaps . . .

We talk a little more about the lives and loves of hedgehogs and I outline to Hugh in more detail my slight misgivings about them, and their popular status in the hearts and minds of wildlife-loving Brits. 'Potter and her ilk, they lied to us!' I joke. 'She's not a washer-woman! She's as sexed up as the rest of 'em!' Hamming it up even more, I really start to plead my case: 'Potter and her lot spun their own version of reality to tell human stories, at the expense of the mind-expanding truth of animal lives and loves!' Hugh says nothing, but grabs his laptop and starts rooting through his personal folders, looking for something. Nodding his head feverishly, he looks through files and files and then stops abruptly. 'Ah-ha!' he says, before turning the screen around to me, to show me what he's found. It's an image, a quote scanned from a medieval-looking text. It reads:

> *That the Hedge-Hog is a mischievous Animal; and particularly, that he sucks Cows, when they are asleep in the Night, and causes their Teats to be sore.*

What? I re-read it once, twice, three times. Hugh cuts the contemplative silence: 'Before Beatrix Potter, most of the mythology and folklore surrounding hedgehogs had them as mysterious, and portents of doom.' He shakes his head. 'I wouldn't ever want us to go back to that time. Never.' To Hugh, Potter saved the hedgehog, at least from *that* kind of misrepresentation. 'Without the hedgehog in its current popular incarnation, where would we be?' he says earnestly. 'I'm immensely proud of how far it's come. Like the robin, the hedgehog is an animal that's become tolerant of people – and that makes people, particularly children, warm to it.' To Hugh this matters enormously. He talks about the sense of connection that hedgehogs bring, and that Potter helped start. He speaks breathlessly, passionately, explaining their cultural success – that hedgehogs are accessible, that they're cute; that childhood encounters with them stick firmly in our mind, and that saving them saves all sorts of other creatures.

'If a hedgehog opens up its ball of prickles and you get to see it, it is, I would argue, undeniably cute. It lends itself to cuteness.' Hugh talks with joy and slowly, in spite of myself, I start to see what he means. As he goes on and on and on, bit by bit, I start to fall in love with the little bastards. As he continues I keep looking at the stuffed hedgehog in the corner. I keep marvelling at those ears. Those beady eyes. And it is cute. God damn it, they *are* cute.

The hedgehog, then. It's come a long way. From portent of doom to bumbling laundry lady to agent of environmental awareness and wildlife appreciation, at least according to Hugh. Yet I was surprised that we appear to know so little about the intricacies of its sex life: that we have little idea of whether that vaginal plug is left by males or made by females; that fewer people haven't commented about the unusual colour of its penis; or that so few people (apart from Hugh) have ever seen, for real, a hedgehog carousel. This, after all, is a national treasure we're talking about here: a widespread (yet sadly declining) species that we can see in our gardens.

In this respect, how do hedgehogs compare with the snakes with which I began this chapter?

Snakes appear to have gone the other way. Once worshipped by the Egyptians (one end of Tutankhamun's tomb is laden with them) as well as the ancient Peruvians, and with star cameos throughout Greek mythology (often as a healer), the snake's reputation undoubtedly took a tumble (in the Western world at least) on that fateful day in the Garden of Eden. Yet some would argue that we appear to know rather more about the sex lives of snakes than those of hedgehogs, since in many species they are often so much less discreet.

The anaconda is perhaps the most famous in terms of its public sex shows, partly a result of it being the bulkiest of all snakes, widely distributed and almost hallowed in terms of the pop-culture that surrounds it (largely to do with the size of the things it can swallow). And, yes, they have also grabbed some plaudits for their sex, which is nothing if not showily licentious. It works like this.

As with many reptiles, females take longer to mature into breeding condition, which means that for every receptive female there may be three or four eager males. In fact, it is not uncommon to see a female lie for three or four weeks in the middle of a heaving, squirming, slow-motion shag-pile of shagging snakes. Sometimes she may entice a dozen males at a time, each of which will attempt to writhe his way into the right position for fertilisation. In competitive circumstances like these it's no surprise that sperm plugs rear their ugly heads once more (possibly similar to those hedgehogs, then), but these male anacondas have another trick up their sleeve (so to speak): they whip out their vestigial legs. Their spurs. They thrust these spurs towards the female's nether regions and begin to scrape them against her furiously. No one appears quite sure why the males do this. It might encourage her to open up her cloaca for one of their hemipenes to enter; similarly, it might annoy her into submission ('what's a hemipene?', you may ask. Well, snakes have two penises because ... well, why not?). It's a breeding ball, and a bloody big one at that. And she, being larger and stronger, probably uses her bulk to ensure the best (and strongest) male is there at the end, fertilising her eggs. She certainly puts them through their paces at this time of year; like a 30-day ride on a slow-motion bucking bronco (maybe that's what the spurs are for?).

What's unclear at the moment is how the males find her at all. They could be tracking her down via a smell-trail, or it's possible that she might broadcast her sexual messages through the air – or both? Even before the term 'pheromone'

THE GREATEST STORY NEVER TOLD

was decided upon in 1959, there were long-held suspicions that smell (or taste) was an adaptation used widely in the animal kingdom, largely hidden from human senses (the ancient Greeks noticed, for instance, that the secretions of a bitch in heat would attract male dogs from miles around). Though sex pheromones can have a number of functions (including advertising male quality), they can be a useful adaptive tool for species that roam over wide areas, often in the dark or through undergrowth. For snakes, therefore, they are integral to finding one another.

I began this chapter with the adder (or northern viper – it is the only snake found north of the Arctic Circle), for it has one of the more unusual sexual rituals I have witnessed, and it is something you can see for yourself, with a little patience and a certain fleetness of foot. In early spring, males usually appear from out of hibernation ahead of females, their basking bodies littering exposed banks near their hibernation sites. They take the time to get their gonads as toasty as possible, aiding the production of sperm for the events to come. Adult males have simply one task, and that is to find a receptive female. A common theme, you might think, but for the male adder the challenge is even tougher. Because of the icier northern temperatures of the countries that they inhabit (including Britain), females may reproduce only every two years. This imbalances the ratio of available males to ready-to-breed females. It's a familiar story, one that we've heard a few times now. The chase is on . . .

Males move up to a mile during this period, flicking their tongues in and out to smell for sex pheromones that seep from the female's skin and from her anal glands. In adders, her pheromone trails are directive markers – they may communicate her location, direction and reproductive status. He homes in, but may often find another male has reached her first. Ritualised battles ensue (called, of course, The Dance of the Adders), whereby males wrestle, essentially keeping going until one of them realises that, for now at least, the time might be better spent doing something else, like finding another mate (male adders have what you might call their Rules of Combat tattooed into their genetic make-up; the first rule? 'NO BITING'). Not having hands, and not being able to bite or spit (or eye-gouge or fish-hook, for that matter), male adders do what any self-respecting street fighter would do under such conditions: they push each other around and roll on the floor a lot. Once alone with the object of his affection, the winner of the scrap continues the smelling. He lashes his tongue in and out wildly and flicks his tail in anticipation. This might be the only sex he gets for a long time.

Though I have met people that reel and cringe at the notion of snakes and snake sex, I hope these words might at least ease your concerns, if only marginally. Snakes are the old romantics, having had millions of years to hone their craft. But then, so have we all.

Pheromones are vital in the early stages of sex in snakes. Sea snakes, centipede-eating snakes, slug snakes, cat snakes, grass snakes, anacondas – all would be lost (literally) without them. However, the nitty-gritty of pheromones – the variation in a given snake's abilities to home in on them, and to sense the direction of them, and whether and to what to degree females leave them lying around – is still largely under-studied. For some snake species, particularly those that spend most of their lives writhing around in and underneath the soil surface, the chances of finding a mate are low at the best of times. And for these species, that familiar hat-tip to rotifer-kind re-emerges: asexuality. Females give birth to females – genetic clones, in other words. I refer here to the delightfully named flowerpot snake (also known as the Brahminy blind snake), which has found its way around the world via, of all things, the intercontinental garden trade. It looks like an earthworm. It lives like an earthworm. It, of course, has sex like some earthworms – asexually. Even though it has found its way across the globe (it is probably the world's most widespread snake), very little is known about where those pesky males have got to. Maybe they sprout up every now and then? Maybe they're genuinely lost forever, another lineage that went it alone, but which (as we saw in Chapter 7) may later come to regret it? The jury is still out.

I've never yet managed to talk anyone out of a snake phobia, and I don't expect these words to help, but it is gratifying to get a chance to write about the sex lives of creatures we often avoid or revile. Snakes are as progressive or evolutionarily complex or sexually remarkable as anything Beatrix Potter could throw society's way. I'll argue this to my death. The same goes for each and every one of the creatures penned by Potter. It matters not what you think of them, or how enraged or enraptured they make you, they all have bigger things on their mind.

There is one animal among Potter's menagerie that it seems fitting to focus on finally within this chapter, because they, like snakes, can fill us with fear and a shrill terror that far outweighs their diminutive size. I wanted to remind you of their sex lives because I have come to rather love them. I am referring, of course, to mice.

Mice are like us, except in many ways they are better; their dangly testicles can be retracted, for instance, and they have five pairs of nipples. And they are also immensely social little things, and mightily adaptable, too. Potter had them covered, of course. Far from making them into washer-women or offering them up brave encounters with farmers, she peddled a story arguably closer to that of the natural world. She had mice down as vandals. In Potter's *The Tale of Two Bad Mice*, two mice trash a child's dollhouse, smashing dishes, tearing down the bolsters and throwing dolls' clothes out of the windows (actions for which later the mice later atone). But she fails to mention what they got up to in the bedroom of that doll's house, possibly for good reason. For mice are renowned for their ability to breed beyond belief.

That's the story, right? Well, here's the reality – as best we understand it.

Sure, mice can breed – when times are good they can end up with more than a hundred offspring a year (they can get to it from about five or six weeks old, potentially having 10 litters per year, each litter containing up to 14 young.). But to label them as simply 'good breeders' does them a disservice. For the secret of the house mouse's planet-wide success isn't in its sex but rather in the *plasticity* of its sexual behaviour.

House mouse society (if we can label it as such) can be surprisingly malleable, depending on where they live and how much food or space is available. Mice that have a handy food resource nearby (in your kitchen, for example, where food may abound) are the ones more likely to become preoccupied with how they share space. Their home range shrinks and, as the frequency with which they trample upon one another's territory increases, so tempers flare. Interactions between mice in these circumstances become highly charged. So frequent are aggressive bouts that their populations become hierarchical – they have to, otherwise everyone would end up murdering everyone else, I guess. This hierarchy keeps most interactions between mice from escalating into chaos. Males certainly get aggressive with one another, an adaptation to keep others in check without risking serious injury, but they also become intensely watchful of one another, ready to eject any sneaky males entering their patch. However, the females in these cramped populations become rather accepting of one another. They don't worry too much about sharing a male, as long as he's of good quality and can defend the territory ably. With an abundance of food and protected from infanticide by the more territorial males, they condone living together in little groups (often made up of related individuals), and they even nurse and look after one another's offspring. They become, in essence, communal breeders.

Compare this with their more isolated cousins in the country. These 'non-commensal' mice live more of a roguish, wild existence. With sometimes only a handful of food or water resources to exploit, males must roam larger territories at any given time to survive. For them, encounters with other males (far from being commonplace) are an event: all-out war. In non-commensal populations, interactions between females become aggressive, too. After all, if food is scarce, who in their right mind would possibly *want* to share? They become protective of everything because there's less to go around: less food, less water, less shelter, less sex.

Commensal or non-commensal, the sex life of the mouse fits the environment in which it lives; it shapes itself around whatever the world throws its way. Mice are natural expressions of one of Darwin's most famous misquotes: 'It is not the strongest of the species that survives, nor the most intelligent. It is the one that is most adaptable to change.' Mice are adaptable both ecologically and sexually. It's one of the many reasons they regularly out-sex you and everyone you know. But there are other aspects of mouse sex that I rather like. For example, the males 'sing' (emitting ultrasonic calls in the 30–110kHz range) and, amazingly, female pheromones induce them to do so. And these songs have variety, too, leading some to compare them, predictably, to those of birds. Adaptable, thoughtful, protective and, well yes, a bit fighty, they are anything but the mindless vandals that Beatrix Potter describes.

I have something to admit as we approach the end of this chapter. I had hoped when visiting Hugh that I'd go on to write a 'take-down' piece about the hedgehog. That was my intention. This chapter was to be an exploration of the sexual history of that outwardly innocent Mrs Tiggy-Winkle and her species. I'm not proud, but it's true. I wanted to sully the hedgehog's pudsy, fuddlesome reputation and discover that the females were somehow harlots and the males were skulduggerous peddlers of filth. Yet . . . no.

Thankfully, I saw sense. And it was when Hugh showed me those haunting medieval lines about the mischievous hedgehog suckling at the teats of cows during the night that I realised the error of my ways.

Nature's stories are so wonderfully rich, and made more incredible by their truth (or the thrill in getting as near as we can to such truth). Snakes, hedgehogs, mice − three animals that we all feel we know so well. Three animals that Western society brackets neatly as fearsome, friendly, pestilent. Yet the story of their sex lives is as rich and bountiful and interesting as anything Beatrix Potter or Disney could ever have come up with. Each a story with a narrative arc, with rivalry and intrigue; with a search at its core, with twists and turns and often a sting (or a spur) in its tail. But the best thing of all is surely that this is a story dreamed up by the universe, a culmination of mindless acts of success and failure, a chain of successful mistakes; a thoughtless process that has led to a simple tale told a million different ways, in a million different styles with the same recurrent dialects and the same romantic dialogue. A conversation between two sexes. And all of it, every story, narrated by silence. Until those early scientists declared humanity's place as centre stage, and told their stories from the human viewpoint. And hedgehogs are part of that tale.

I guess I came to love hedgehogs. They have purpose. As much purpose as anything, I suppose. Wasps, snakes, clownfish, spiders. All are peacocks, all pandas. All as equal as one another. All as adept. All made not in Man's image, but in their own. Theirs, all too often, the greatest stories never told.

CHAPTER THIRTEEN

Sex: The Arena Tour

The following news report is from the *Eastern Morning News* from 1897.

> *The Humber Conservancy Commissioners have not yet solved the difficult problem of how to rid Reed's Island of the plague of rats that now infect it, nor are they – the most ready means, the flooding of the island having failed – likely to do so in a hurry. What a few years ago was a splendid pasture land, and sustained thousands of sheep in its rich verdure, is now the home of myriads of rats. It is burrowed from end to end, and so densely populated is this habitat of the rodent that it is said that it is almost impossible to put your foot down without standing upon a rat hole. It is only about a year ago that the rats got*

*the upper hand, and the commissioners have of late been very much
exercised as to the eviction of their unwelcome tenants. It was decided
to cut openings in the banks which surround the island, and thus
let in the Humber waters at spring tide, with a view to drowning
master rat and his numerous family. The openings having been made
at considerable expense, the water was let in a week ago last night,
but not with the result anticipated. As the water advanced the rats
fled from their holes in tens and hundreds of thousands, and made for
the banks which remained high and dry. The screeching and snarling
of the rats as they fought for foothold beggared description. Many
were, doubtless, drowned by the inundation, but, being for the most
part expert swimmers, the impression made in the numbers of the
great army was practically nil. Adjudging rightly that a day's sport
might be had shooting the vermin, a party of gentlemen went down
on Saturday . . . Hundreds of rats succumbed to the firing of these
gentlemen, and it was manifest that extraordinary measures will have
to be taken to rid the island of the pest. The inundation seems to have
done little good, and to shoot them down would be an impossibility.*

The funny thing about this report is that they weren't
actually rats. At the time they were thought to be rats, of
course, but soon after the article was published it was
revealed that, no, these weren't rats after all but something
else entirely: water voles. Semi-aquatic, wide-faced, chubby-
cheeked rat cousins, with short fuzzy ears. And the reason
for their incredibly high numbers? The island, and others
near it, had been colonised by an invasive plant: 'a dense
crop of some succulent weed, supposed to have been
imported from a distance'. The water voles loved it and, in
the absence of predators, had their fill. Their sex then filled
the island.

Water voles, as you may or may not know, are yet another
species sliding toward extinction. They've taken a battering.
In Britain the species has declined by 20 per cent in the time
it has taken me to write this book. In fact, their survival in
this country now depends, in part, on captive breeding and

reintroductions back into suitable habitats. I've been hearing about them a lot in recent weeks courtesy of a host of articles in the press. At one point, while nursing a slight bit of man-flu, hot, sweaty and restless in bed, I listened to a Radio 4 programme about them. That was when it hit me. Where is conservation going? Will more and more creatures teetering on the cliff face of extinction be subjected to sex in artificial enclosures for later release in the wild? Isn't that a bit weird? A bit sad? Will all such future efforts depend on scientists watching sex acts behind two-way mirrors with their fingers crossed? I had a momentary feeling of angst; a haranguing uncertainty about where evolution was taking us. Humans, the first animals in the history of Earth that seek, often inexplicably, to help other animals have sex in buildings often purely for the thrill of knowing that they may one day continue to have successful sex in the wild without them. What a strange species we are becoming, wouldn't you agree?

These peculiar thoughts occupied my brain for a number of weeks as summer turned to autumn. It became a topic that I knew warranted further thought. And then, I got an invitation. An invitation to something very special.

*

It's early and I'm sitting in my car in a layby on the A11 to Norwich. Someone has sent me a video on my phone, to ready myself for today's meeting with a designer of sex arenas. The windows are suitably fogged up for such a deed and I have the volume on my phone turned up loud. The little screen opens with a still pond's surface. It pans slowly left and then up and along the sumptuous legs of a horrendously large spider, which is playfully tickling the water's surface. 'This is a male fen raft spider,' says the narrator. She speaks like a sentient robot, a polite grace to every syllable. 'And he is looking for a mate . . .' Piano music plays lightly in the background as the camera pans across his long legs. They look like those of a male ballet dancer, gracefully pointy, arched together; each packed with zinging levels of potential energy. His body sits proudly atop those legs, shaped like a Formula One car chassis (complete with decorative white stripes down the side). Like headlights, a set of large eyes pokes out of the top, and the beast's mouthparts jiggle in excitement as he continues to delicately tap the surface of the water. I don't have a fear of spiders exactly, but neither would I say I want to actively look at them close-up for long periods of time, cramped in a small car in a layby on a dual carriageway. Yet this one, delicately making waves on the pond surface with its feet, has me hooked. It has character. It is an evolutionary concept-car for spider-lovers.

The video continues. The camera pans across the pond and refocuses on a bigger, chunkier spider standing with its back to the male. The narrator continues: 'And this is who he is signalling to. It's a female.' Her eyes are like portholes. There is a sheen to her, as with an aquatic mammal. A pelt. She's quite sexy. According to my narrator, the female spider has been walking around leaving behind a trail of silk that communicates to local males her sexual history and her readiness to mate. The camera zooms up to her rear end, and her spinnerets glisten in the studio lights, a thread being pulled out like a line of ticker-tape, with a message of sex for the male spider that roams around behind her. 'The male will follow this line wherever she leads.'

Suddenly the piano backing-track is edged up. It plays the notes with a kind of dramatic potency, a sign of what is to come. She notices him. It's game on. He comes closer, nerves jangling. He stops. Shudders. Edges forward. Stops. He comes closer again, slowly this time. Measured. She stays still now, and lets him approach. Within a minute or so, they are face to face, and the piano drops a key. They stare at one another for a few moments before the narrator speaks softly: 'If she approves of his dance, she will allow him to approach her . . . But he must be careful.' She pauses. 'Female raft spiders, on occasion, have been known to kill males.'

The fen raft spider. Arguably the most gracious and regal of all British spiders. And, like water voles (and pandas), they've had it tough. Remarkably so. While it's likely that at one point they existed over whole networks of lowland Britain, now they remain at just three sites. Three. And that's it. Three populations separated by hundreds of miles. This is often what it looks like just before a species goes extinct; little islands of populations, survivors of a largely sunken ship, clinging on to their life-rafts. Dead spiders walking.

But the fen raft spider has several things going its way – it is big and noticeable and showy. In fact, it is one of the biggest spiders in Britain, and for this reason people know about it and many rather like it (everyone loves a title, after all). As well as being big, as spiders go it is highly unusual. A water-walking ambush predator, the fen raft spider uses its front legs to sense ripples on the surface or underwater, and lunges at unsuspecting prey that drifts too close. Imagine a crocodile that, instead of ambushing land predators approaching the water, could walk on the surface and ambush the creatures that lived beneath. Well, it's a bit like that, but with a spider.

In 1999, the fen raft spider became the focus for an ambitious Species Action Plan in England and Wales, a project that sought to reintroduce it to parts of its former range. Releasing batches of healthy young spiders, some of which it was hoped would later breed, was a key part of the plan. But there was just one problem – how do they breed? In the decade that followed, the fen raft spider, a normally secretive species about which very little was known, was to have a magnifying glass held up to its intimate sexual habits.

I take a second to wipe the screen with my sleeve as the action hots up. The video on my phone, by wildlife film-maker James Dunbar, shows how far the project has come.

The video is perhaps the first in the world to detail so clearly to a global audience the sex life of the fen raft spider. I sit quietly, watching my phone as the mating ritual continues. Their outstretched legs touch, tip to tip. The piano strikes a firm chord. The narrator almost sounds as if she's smiling: 'She is willing,' she says, letting the sentence hang in the air.

The male spider inches closer, but still approaches her as if she's a booby trap that could go off at any minute. He gets nice and close and then ... what the ... ? He suddenly takes to bobbing up and down on his legs, like a wind-up jack-in-the-box. If spiders have arses, then he is shaking his. It's the spider equivalent of a twerk. She looks a bit confused by this, and continues to quietly watch while his bobbing reaches fever pitch. This is his dance. Without warning he starts interspersing the frantic bobbing with occasional bouts of leg-wagging in her face – as if it's something he's only recently learned, his clincher; his best move. The male looks like a baton twirler on a vibrating treadmill. 'This violent flicking motion signals that he is on the home stretch,' says the narrator. He stops suddenly and assesses the female's reaction, before plucking up the courage to approach her and go for it. It's now or never. The female grows suddenly immobile, then lifts her legs and allows him to climb on top of her. And then? Well, let me tell you. Let me tell you indeed. For at this point, things take a turn for the even more bizarre.

There is suddenly a lot of grappling and rolling, and it's hard to see what, exactly, is happening. This continues for a few moments until they stop, as if to allow the camera to take in the enormity of what's going on. The male has her held on her back, in some sort of bizarre choke-hold. He then moves slowly and calmly, working on her front end like he's defusing a complex bomb. He feels across her body with his two palps (modified male appendages) and then, upon finding her genital opening, plugs in and proceeds to pump in his sperm. There are a few quiet moments when all is calm, and then ... they wrestle. They flip, spin and wriggle and then, as quickly as it started, it's over. She flips herself back up, as if nothing had ever happened. The male turns the other way, takes a few moments to sort out his palps (I'm still amazed at how much cleaning animals give their genitalia after use) and then he's off. 'They will part ways now,' our narrator says, almost wistfully. 'The male will try to mate with as many females as he can, as this is his last summer.' And that's that.

I feel a bit dirty, sitting like this in my car, but it probably isn't the worst sex-act ever seen in this layby. I turn off the phone, put my car keys in the ignition and head off to Norwich, occasionally checking my shoulders, my hair and my neck for horny spiders. For this is the starting point of my journey into a blooming conservation science. The sex arena.

Dunbar's video of those spiders was set in one such place – a lab-based enclosure that has been carefully prepped with the right ingredients to make spiders have sex. These sex arenas are useful to scientists for a host of reasons; by manipulating variables like temperature, habitat condition or sex ratios you can observe the impact that these factors have on mating behaviour or reproductive success, both of which are crucial pieces of information if you want to reverse the fortunes of a species threatened with extinction.

Ian Bedford was the first to answer my call about sex arenas, and he's invited me up to Norwich to the John Innes Centre where he holds the post of Chief Entomologist, a job title that makes my inner eight-year-old swoon. Here he leads research and assists on a host of projects that focus on understanding the sex lives of animals, particularly those whose sex lives occasionally endanger the crops on which we depend.

Ian is a jovial chap who has clearly lived and breathed this place for an age. He shakes my hand with a smile and an almost-giggle, eager to hear about what other animals I've met on my animal-sex journey. Starting life as a technician here, Ian has now cultivated the vital skills of many a modern-day scientist: three-parts natural curiosity, one-part business acumen. We wander down through the hangars before Ian leads me into a dark room, and into his lab. As with many entomological labs I've visited I'm struck immediately by the rich smell of decaying lettuce mixed with a fruity bouquet of recent stick-insect droppings. Hissing cockroaches sit in a plastic tub on my right, alongside some African land snails. Some slugs are in tubs on my left (I don't spot any slug mites). In what looks like an empty Ferrero Rocher box on the side there sit some shed skins from a tarantula. Standard stuff, really. Paperwork litters the lab-tops, and every spare wall space is filled with scientific posters. Above some lockers in front of me, long jars are filled, unsurprisingly, with stick insects.

We head into Ian's meeting room and sit down. I am more than a little disappointed that, so far, I haven't seen a single sex arena anywhere. Ian has other things on his mind, though: 'Do you like slugs?' he asks in all seriousness. 'I'm coming to love them,' I answer. He smiles. 'Let me tell you first of all about my newest project . . .' he says, leaning forward to spill details on what is clearly a recent obsession. The spiders will have to be put on hold, I think.

He outlines his interest in slugs, namely that such creatures are often considered pests. Nearly all slugs are, of course, hermaphrodites. And hermaphrodite slugs – especially the annoying pestilent ones – pose unique problems for some humans, namely because they have twice the potential to procreate, compared to most other animal life forms. Each individual, provided it can hook up with a mate (and many don't even need to do this), can have offspring. That makes them problematic for humankind. Essentially they can go from a population of two or three to a population of ten-thousand-tiddly-stupid in a matter of weeks or months. If you're a farmer, pests like this have the power to bankrupt you.

Ian's new obsession is the Spanish slug, a noted invasive species. Only moments after sitting me down he shows me a cutting from the *Daily Mail*, pulled from the wall of his lab. 'THE FIVE-INCH KILLER SLUGS INVADING BRITAIN' is its headline. 'Millions set to attack our garden plants and vegetables after arriving from Spain.' The article includes a big photo of a slimy mass of them eating a snail. It is gruesome, no doubt about it. Ian now has my full attention.

Incredibly, this entire invasion story starts, of all places, in Ian's garden. 'Last year I noticed there were massive problems with slugs in the gardens around here,' he begins. 'I noticed that all my native slugs were disappearing and being replaced by just one type of slug. It was an odd slug. Really odd.' He pulls out a photo of a large, nondescript orange slug sitting in the middle of what looks like an incredibly enormous dog turd. Ian tells me that it is, indeed, an incredibly enormous dog turd. I express disgust (at the turd), and Ian tells me he had to crop most of the turd out of the photo for fear of upsetting people (he also never refers to 'dog turd', only to 'dog mess', which is rather endearing). I am momentarily grateful for his measured words, though my disgust remains. Call it what you want (or crop it how you like), it is a disgusting image. No doubt about it.

'These slugs are really weird,' he continues. 'Often slug species normally just bumble around and bump into food, but these Spanish slugs, they actually seem to rear up like a snake to sense what's around them. Particularly when they're homing in on dogs' mess.' I mouth a swear word to convey my surprise and the tiny bit of delight I feel at hearing this. They sound like something out of a cartoon or a straight-to-video B-movie horror. There's more, though. 'In the garden I saw them eating all sorts,' Ian goes on. 'You'd see them eating freshly killed mice that my cat had caught; you'd see them going through the garlic and the onions, destroying plants that slugs normally don't go for.' Garlic? Mice? 'Things got so bad locally that many people reported these things coming in through cat flaps and devouring the cat food.' Cat food? Cat flaps? I quietly whisper another expletive. On witnessing these strange slugs, thankfully Ian was on the ball. Suspecting something wasn't quite right, he took samples of the slugs and sent them off to the UK's slug expert, Les Noble at Aberdeen University. Studies of the slugs' genitalia showed them, indeed, to be the Spanish slug *Arion vulgaris*. It was the first such discovery of *A. vulgaris* in Britain. The Spanish slug had arrived.

It is undoubtedly an odd creature. It almost sounds as if it has been dropped onto Earth by a passing spaceship, so little seems to be known about its history and how it is managing to colonise new places so quickly. Thought to be native to southern Spain, it has been moving (slowly, of course) northward and eastward in the last half century. It arrived in Switzerland in 1956; then Italy in 1965; Germany in 1969; Austria in 1971; Belgium 1973; Norway 1988; Finland in 1990; Czech Republic in 1991 . . . you get the idea. It is moving at about the same speed through Europe as EuroPop is retreating. A very, very, very slow crime wave. And now it's here. In Norwich.

The *Daily Mail* headline had referred to them as 'killer slugs', yet . . . I don't know . . . they seem nasty-looking, I agree, but I can't imagine being chased down the street by one. 'No,' Ian chuckles, then gazes sternly my way. 'But it's more that they are an unknown quantity.' He looks suddenly rather grave. 'We know almost nothing about the impact they might have, and how we might stop them.' They caused a colossal amount of damage to local fields in recent years, with many farmers having to sow their oilseed rape three times to overcome the enterprising molluscs. 'We were also contacted by numerous gardeners. One lady was killing three and a half thousand slugs in a month. Another chap, four thousand slugs.' I wonder momentarily how one disposes of four thousand dead slugs, then think better of asking. But just why is this slug so successful? Ian explains it expertly, referring to a kind of perfect storm for Spanish slug sex. Essentially it is a slug that's moving (care largely of the plant trade) away from its native world – a world where sexual resources are scarce – into a land of milk and honey; a world where the key ingredients for sex are everywhere. Rain, mildness, food: Norwich.

The Spanish slug is used to arid environments. Over millions of years it has adapted to a world of harsh dry summers and cold dry winters. Its slimy skin protects it from drying out; it has tough-shelled eggs able to survive the worst of the weather; its senses have been honed to track down food items, which are often few and far between in dry environments. Like vultures (another scavenger of arid regions) its sense of smell is incredible. That, then, is the world in which these slugs have evolved. In the middle of these extremes is a wet season – a few weeks, really – but it's the time when the slugs can move around freely, finding other slugs with which to mate. 'Now, take one of these slugs and deposit them in a mild wet country and you can see the problem,' Ian tells me. Here in Britain, the ingredients for sex are everywhere: wet weather, and plenty of food for growing new bodies. Those hardy eggs, and the fact that many native predators apparently avoid the Spanish slug (too slimy), mean recruitment of baby slugs into the adult population is high. And this, of course, means yet more sex is on the cards. 'The problem is compounded by the fact that you only need one slug to start the whole thing off.' Ian pulls out another handful of photos from his file. 'Being hermaphroditic self-fertilisers, whole populations can start from one individual that finds its way through the borders in a pot-plant.' It is a profligate, self-fertilising, big-nosed battle-axe of a slug. And, according to Ian, it's coming to a neighbourhood near you.

I joke with Ian a little at this point, if only to mask the imminent terror I have of the new world order, a world where garlic-addled slugs steal food from cats right in front of our noses. Ian puts my mind at rest. 'We think we might be able to do something, we've just to work out what that might be.' Sex is the problem, sure, but is it really feasible for us to control its sex life? Ian is quick to respond. 'You can't stop them once they've got in – it's true.' He clasps his hands together like a politician. 'They're out there; we'll never control them totally. But where this slug is a problem, we have to find some way of dealing with it.' And what might that be? 'That's what we aim to find out,' Ian answers. Sex arenas will be the key.

Traditional techniques for putting a stop to slug sex can be effective, but they may be riddled with unwanted side-effects. For instance, molluscicides (slug poisons), when sprayed, can end up in water-courses where they kill off native snails (with this possibly contributing to algal blooms). Likewise, garden slug pellets have been associated with frog deaths in the past. Introducing nematode parasites into populations of unwanted invertebrates like slugs is another option, though whether this is feasible for the Spanish slug (which appears quite hardy with regard to these worms) is anyone's guess.

How exactly can we take on animals like these that we need to control? In the last 10 years the answer has become more and more obvious. Now, more than ever, scientists are stocking the metaphorical gun-rack and their bull's-eye is sex. Of the techniques used to stymie the sex lives of bothersome bugs, one recent fad has been the breeding and releasing of sterile males into wild populations. It's an interesting idea. You flood populations with them and watch as the infertile males mate with the females, which then lay dud eggs. Result: the population plummets. Mosquitos are pest species upon which this technique has been tried, not once or twice but a number of times. In theory it sounds great, with the main perk being that no other species get caught in the cross-fire. On the downside, though, in practice the effects are unlikely to be lasting – applications of sterile males have to be repeated as sexually successful females recolonise. It's a band-aid, basically.

Pheromones may be another tool that scientists are turning toward to stop the onslaught of invasives. For example, scientists in the United States have been toying with 7α, 12α, 24-trihydroxy-5α-cholan-3-one-24-sulphate; that extraordinary mouthful is actually a manufactured sex pheromone that has the power to attract female parasitic lampreys into pheromone-baited cages, where they won't breed, multiply and drink the blood of other (prettier) fish in the wild. If the trials continue to be successful, it could be good news for the fishermen of North America's Great Lakes, many of whom detest the sight of a hooked game-fish with an ectoparasitic lamprey dangling off its body. And this would be great for the game-fish too, of course, particularly the lake trout, an apex predator whose populations collapsed as a result of the lamprey's accidental introduction to the lakes in the 1930s and 1940s.

'Could pheromones work on the Spanish slugs?' I ask Ian. 'Not sure . . . you see, the situation is complicated by the fact that these are hermaphrodites we're talking about,' he answers. 'This is why arena trials have such value.' His project will look at the genetics of the slug, how it breeds and how it might be controlled. Slug sex arenas will be a key part of the project.

I choose this moment to come clean with Ian on something. I had imagined that these sex arenas would be everywhere in Ian's lab. Large plastic gladiatorial colosseums, with entrances at both ends – one for the male, one for the female – surrounded by flocks of scientists watching, scribbling down notes. I had imagined Ian, sitting like Caesar, overseeing everything. But, shucks, no. Turns out that, in reality, sex arenas (for invertebrates at least) are often little more than plastic trays, lined with a spattering of bits and bobs in which to forage or hide. It'll likely be this way for the slugs.

The conversation moves from techniques to stop animals having sex (Spanish slugs) to techniques for encouraging animals to have sex (fen raft spiders). We talk about the fen raft spider sex arenas that were once located here. Ian makes them sound rather simplistic. The fen raft spider containers were half-filled with water, he tells me, in which sat a few upturned plastic tubs alongside a smattering of pond vegetation, and that's about it. 'From what I can remember, it took a bit of time to get the fen raft spiders to breed,' says Ian. 'A bit of trial and error at first, certainly, as we tried to find an environment in which the male and the female were in a comfortable state.' He gets a little frustrated with himself at not being able to remember all of the details. 'We realised that one of the things that really seemed to matter was giving space to the males, allowing them the time and a good location from which to drum.' Ah, the drumming. I think back to that layby on the A11. The female raft spider sitting with her front legs in the water, feeling for his drumming vibrations on the surface – it was these that she used, at first, to monitor his advances.

'We knew we were succeeding when the female would approach him, and he'd manage to flip her over,' says Ian, reliving it in his mind's eye. 'From what I remember, that bit was always a fairly rapid thing.'

The next bit wasn't. Though breeding was apparently easy to reproduce in a sex arena, the next stage proved a little more difficult. Sometimes females would ditch their eggs for no apparent reason, or eggs would mysteriously fail to hatch. But these were teething problems; problems that have now been tackled by years of tinkering with the experimental approach. According to Ian, many of the lessons learned about fen raft spider sex came from sex arenas here at the John Innes Centre (scientists now expect to rear over 90 per cent of the fen raft spiderlings in a given batch of eggs – an incredible statistic).

At this point Ian gives me a stern bit of advice. 'Jules, if you're going to be writing about fen raft spiders,' he says, 'then you need to speak to Helen Smith. She organised and championed the whole thing. Without her we'd know almost nothing about them, and the whole reintroduction process would have faltered.'

I duly note this down, before shaking Ian's hand and saying goodbye. I wish him the best of luck with his Spanish slugs and make my way back to my car, stopping to examine each and every slug I see on the way across the car park.

Later that evening, I attempt to track down and make contact with Helen, to learn more about the recent objects of my affection.

*

After 10 months of immersing myself in the sex life of animals, I've started to appreciate that the best, most experienced animal scientists are those it's hardest to get on the end of a phone. They might be in a basement operating an MRI scanner, or in deepest darkest Peru, or in a submarine. They are the scientists you want. They're the ones with the best quotes and the most authoritative voices. Well, trying to get hold of Helen Smith was a bit like that.

She spends much of her time in the Fens, where the skies are big and the number of signal bars on your phone is small. And she's proven hard to track down. After a few stolen conversations on the phone, though, we arrange to meet, and in the best, most wonderful, circumstances – releasing baby captive-bred fen raft spiders into the wild. I can barely contain myself.

Ian was right, Helen Smith is indeed an integral component of the fen raft spider breeding success story, the fulcrum upon which every action has seemed to pivot. You'd never know this, though, from speaking to her. Helen tends to retreat from such praise, this being a partnership project involving hundreds of people and organisations (including almost a dozen British zoos, which now take care of the captive rearing of the fen raft spiderlings each year).

Her story, and the story of the captive-breeding programme's early days, is worth telling, mainly because it begins so humbly – namely in Helen's kitchen. 'What I would say about rearing the spiders at home is that it gave me privileged access to them, day and night, throughout their breeding season,' she told me in one of our early email correspondences. 'I was able, routinely, to witness and record their courtship, the construction of their fabulous silk egg sacs and the eventual emergence of their young – all very rarely seen in the wild.'

Rearing such spiders from tiny spiderlings is tough, she told me, not least because each demands its own personal supply of flies. 'I spent much of my summer hanging around compost bins and sweep netting over pony dung. Keeping them supplied with tiny flies was a full-time job.'

Once the spider's sex life was more fully understood, captive breeding and rearing could take place, and the release project is now in full swing. At the time of writing this, fen raft spiderlings have been released at three new sites in Suffolk and Norfolk. Wild breeding has since been observed at each site, through surveys and the monitoring of 'nursery-webs' (the silken meshes of webs spun by the broody female in which her babies will seek safety before dispersing).

When I finally get to catch up with Helen, it's in the car park of a more recent release site, the RSPB reserve of Strumpshaw Fen in Norfolk. It is spiderling release day. As we make small talk she flips open the boot of her car and allows me to ogle her precious cargo – three big plastic tubs filled with test-tubes, in each of which sits a single spider. My mind reels at this sight. There must be 1,500 or so. I itch only slightly. Though they are no more than babies, they seem bigger than I imagined, with a body about the size of a ladybird (a ladybird on long stilts). They look soft, though, not spiky or armoured like a beetle, or jagged like a garden spider. Modest, light, water-filled almost (which, I guess, they kind of are), and with a slightly velvety texture. As with the female I saw on the video earlier, they have a waxy fur to them, like otters or seals. Nearly all sit poised and still in their test-tubes. They await their fate patiently.

Helen outlines the plan for the day, which is refreshingly simple: find the release site, find a suitable microhabitat within the release site and unplug each test-tube to let the little spiders out, as and when each is ready. I had imagined we'd be shaking them out of the test-tubes, but no: Helen tells me that we only open the lids; they choose when to emerge (she comes back to collect the empty test-tubes in a day or two). Jokingly, I am told that I should take a tissue or two: apparently newcomers like me are sometimes prone to emotional tears at such a climatic event as this. I laugh nervously, wondering if she's being serious. 'But, well, how about you?' I ask Helen. 'After all these years do you still get emotional about this?' Helen plays it cool. 'I used to,' she smiles. 'In the first years particularly, having had all these little spiders in the kitchen and in that very cosseted environment, and fed them all individually, and then taking them all out, releasing them into the wild somewhere like this . . .' She pauses. 'I guess I thought – oh crikey, my babies are going!' I laugh. 'And do you still get this feeling?' She chews on this for a moment. 'I get that feeling now from seeing the nursery-webs on the new sites – the sign of new spiders breeding in the wild. That's what it's all about. It's the end product.'

Helen is optimistic about how the releases are going. At each of the three release sites, new nursery-webs are present, in encouraging numbers, but she remains guarded and refrains from getting cocky. 'It's gone better than I could have hoped at this stage,' is the most positive statement I can get out of her. Helen knows that there is still a great deal riding on such a project, not least the risk of a freak weather incident like a flood or a sea-wall breach that could set the work back immensely. Still, it's looking good so far.

We set out across the reserve. After a short bit of walking we find ourselves at the release site, a series of tussocky fields blockaded by a network of ditches. The water in each ditch is crystal clear and great hordes of whirligig beetles circle and fizz, driven wild by the sudden appearance of an autumnal sun. The browning tips of the water soldiers protrude from the surface, providing handy perches to passing damselflies and caddisflies and the occasional hulking dragonfly. We put down our boxes of spider-filled test-tubes and I turn to meet Tim Strudwick, the RSPB's site manager. Tim is a friendly chap who's clearly proud that his site is playing host to such an illustrious spider as this. And then we set to work. A couple of hours seem to pass in no time at all, with us taking lids off test-tubes and carefully allowing the baby spiderlings to enter their new world, this ditch in the middle of Norfolk. Surprisingly, to me at least, each spider appears to have its own personality. When the lids are removed some spiderlings refuse to budge, happy and safe in the humidity of the test-tube; others are totally different – they make a cautious approach to the edge and stand suddenly still, weighing up the relative risk of life out there in the big wide world. One or two spiderlings clamber quickly out of the tubes, and dance their way through the undergrowth, among the dead leaves and down to the bank side – to their new life in the wild. Most won't survive. Some will . . . and then next summer, if all goes well, they will breed.

The evidence of wild sex from previously released spiderlings is apparent all over the place; along the way we see plenty of empty fen raft spider nursery-webs, the spiderlings gone and the mothers now dead. The low autumn sun makes them almost glisten, like Olympic torches held high above a stadium; their stadium, the ditch. Their new arena. It is a lovely feeling being part of it, if only for a couple of hours. I don't get emotional exactly (I don't have tissues, for a start), but I blather about my gratitude to

Helen and Tim throughout. Another hour passes and then we're done. The crates, once filled with spiderlings and test-tubes, now house nothing but test-tube lids. Hundreds of test-tube lids. It is time to head back.

Just as we're about to start walking, I hear a slight gasp. Tim and Helen pull out their close-focus binoculars and peer at what looks like another empty nursery-web. 'There – can you see it?' I can't. I lean in closer and allow my eyes to refocus. Nope. Helen hands me her binoculars and I'm met by the astonishing image of this, my first wild adult fen raft spider, a late female sitting atop her nursery-web. She looks majestic, facing the wind, like a conquering hero (which she is, I guess). 'It's likely she's on her last legs,' says Helen, deadpan. Her babies have formed a huddle below her, deep within the safety of the nursery-web she has woven for them. It's hard to believe that she was raised in a zoo, a lab product. Yet here she is, her life's work complete, her babies starting their own, completely wild, journey. A journey toward their own sex. Wild sex. All that the conservationists have ever wanted. All that they ever dreamed of.

In that moment, I can see why Helen and her colleagues find this a little emotional. Releasing thousands of these spiders, after being such a big part of their lives throughout infancy? It's strangely touching, wouldn't you agree? Suddenly the fact that they are spiders doesn't seem to matter. Life's beauty lies not in the species, but in the journey. Spiders that are the products of a sex act orchestrated not by nature, but by (to them) funny-looking primates like Helen and Ian – primates with a love of life, making lives, hopefully, of love, or a kind of love. 1,500 lives, to be exact – and how many more will come from them?

I am not sure what you feel about wildlife conservation, whether it is money well spent or not. I suspect you might feel, like me, that often it is. But what this October visit showed me more than anything were the nuts and bolts of the conservation idyll. Sex: the bottom line, the currency of saving or killing off a given species. Sex in artificial enclosures, the fiscal stimulus. Sex in the wild, the free market. And like economies, it's sink or swim for the players involved – some animals quick off their feet, sparky and better suited it seems to the conservation model, like those fen raft spiders. Some sluggish, slow to start and forever, like my pandas, in need of subsidies that may last for decades, maybe longer. And like the water voles at the start of this chapter, it's an economy that will never, ever move away from that simplest of ecological or biological models: a never-ending boom and bust, regulated by resource availability and resource stability, itself increasingly regulated by us. Boom. Bust. Boom. Bust . . .

CHAPTER FOURTEEN
My Chemical Romance

Heavily as stones they fall, fall to the tops of the firs where they suddenly sprout wings, become birds and then light feather rags that the storm seizes and whirls out of my line of vision, more rapidly than they were borne into it.

Konrad Lorenz, *King Solomon's Ring*

It's early in the morning. I'm in the car again. Even though the clocks have gone back an hour I'm still having trouble waking myself up. The two black coffees did nothing for me. The sun has barely risen behind the streaky November sky. When it happens, it's sudden: WHAP! There's a knock against the bumper. I look in the wing mirror and there is a big black bird lying in the road, shaking and writhing wildly, like a fly-tipped bin-liner catching a gust. 'Jeez,' I think, parking up the car. What I see as I get out and walk along the road verge makes me feel terrible, rotten right down to my core. It is a jackdaw, incredibly close to being dead. Bloody hell. I'm suddenly unsure of what exactly it is that you should do in situations like this. Do I take it home and attempt to nurse it back to health? Do I put it out of its misery? If so, how exactly do you bump off a jackdaw? I stand in silence, a mournful look on my face. Some cars drive past me, no doubt wondering what the hell I'm doing. And then, at that point, I notice something else. Something that is to trouble me deeply.

That half-dead jackdaw is joined by another. A living one. It hops out of the bushes and nervously jigs forward, cocking its head slightly, eyeing up its mate while keeping one jewel-like eye on my approach. A couple of gruff chokes come from the injured jackdaw, as if it were close to drawing its final breaths. This is the last thing that this dying bird is likely to see: its lifelong mate, standing there, head cocked, as if digesting the thought of facing the future alone. The healthy jackdaw comes closer. Inspecting its fallen partner a little, it clucks out some *mawks* and *kaks*. And then, it is all over. Out of nowhere a car fires round the corner and finishes the dying bird off. As the car disappears around the corner there is a moment where I stand with that living bird. In silence we stare at the dead body.

Now, I know what you're thinking. I feel a little foolish for attributing to this jackdaw human emotions that only I was likely to be feeling. We will, of course, never know what it truly felt, but I wondered at that point whether one day we might at least get a little closer. A little bit nearer to putting scientific language to the attachments that animals feel for one another. I hung around for a while after the incident, waiting to see what would happen to the lone survivor without its mate. This proved hard to do. It retreated into the bushes and refused to come out, at least not until I walked back to my car. I sloped backwards, away from the crime scene. There, I sat and watched in the rear-view mirror as this jackdaw came out to inspect its fallen partner, the one with whom it may have nested for years and years, and the one with whom he or she had taught their offspring how to be flying, looping, soaring new jackdaws. Inevitably I ask myself, 'Did it know love?'

You are right to roll your eyes at me. I mean, here I am getting all soppy over a dead jackdaw. After all, humans lose partners, children, friends, family – we starve, we get diseases, we're abused, we're unfairly imprisoned, we die in wars. Shouldn't our sorrow be saved for our fellow man or woman? After all, we DEFINITELY feel pain, and a pain we can understand, too. And yes, I'd largely agree with you. But there's something about those jackdaws. There's something mystical, hidden; something complex about them. Konrad Lorenz spent years studying the jackdaws that lived near his Austrian manor. I'm starting to see the attraction. Perhaps Lorenz was onto something . . .

In recent years I've become a big fan of jackdaws. From our bedroom window there were seven nests this year within eyeshot, plus one in earshot (the one in our chimney). The same pair (I think) has nested there each year, and each spring I wake to the telltale sound of sticks being dropped from above, like someone rustling crisp packets behind the bedroom wall.

Jackdaws are fascinating birds for a host of reasons, not least because they are monogamous. Now, monogamy isn't a term I've used very often in this book, largely because it is as unlikely a reproductive strategy as you could expect to have ever evolved on Earth. It's rare. It's wasteful. It makes little sense for most creatures. But jackdaws like it. And they do monogamy properly, too. They are MONOGAMOUS. Truly sexually monogamous. Or rather, no one has ever proved that they're anything other than completely monogamous. There is no evidence of extra-pair copulations, or any other funny business like that. Nothing. Males and females pair up and remain faithful for years on end. And that's it. The end. Happily ever after. They make nests, they raise broods of chicks, they forage together and they roost together in the colder months. They are likely to be the most sexually faithful animal within 100 metres of you right now. For that, depending on your feelings about monogamy, they deserve your utmost respect.

Lorenz's soft spot for jackdaws was legendary. He was among the first to expose their mental capacities, showing, for instance, that they ranked themselves in linear hierarchical groups, like apes. He appreciated that they were quick learners, and that they could pass on information, such as who or what to fear, to their offspring. Yet we've since discovered more about them. There is evidence that jackdaws possess a kind of 'theory of mind' (complex thought that can be summarised in the sentence 'I think that you think that x is y'). Others have even hinted at possible consciousness in jackdaws, that cherished grail in all its manifold and nebulous grace. Either way, there's a reason many who study jackdaws (and other corvids) refer to them as 'feathered apes'. They're brainy birds indeed.

For an animal so cognitively advanced and monogamous, it seems fair to ask questions about love at this point. 'Love?' you ask, agog that I even dare bring it up. Yes, love. Now, you may be uneasy at this. After all, this is a book based on science, a method of thought that prides itself on its measurable windows on truth. To most people, love is anything but measurable. Most people in love can barely find the words to express it. Philosophers, poets, musicians and a generation of scholars have wrestled and struggled for centuries to tie it down. 'What place does it have in biology?' you might ask. I'd say it has more place now than it's ever had. Love is adaptive, after all – it bonds pairs together (even if only for a short time), influencing the reproductive output of the individuals on which it acts. And, strange as it sounds, it can be measured too.

For scientists to agree on love, what they needed was a definition and something measurable, based on what, exactly, love does to the human body. What they settled on was hormones. Measurable brain hormones.

Now, you're likely to be uncomfortable with this idea. I know I am. For referring to love in this way implies somehow that, if you've ever been in love, it came courtesy of chemicals. It threatens to cheapen that which you felt. It implies that your happiest, most passionate encounters with a partner were nothing but emergent phenomena that sprang from nothing more than a molecule-addled brain. But, in essence, that's what they were.

Human love means different things to different people, of course, but few would fail to agree on the basics. The commonalities. The behaviours and emotions we share when loved-up. Here's my stab at trying to describe it to an alien entity, not of this Earth.

Being around the object of your affection feels good. They occupy your thoughts, and you crave their touch, even if it's nothing more than brushing past them. It feels nice. You think of them more often than almost anything else in your life. The thought of them with anyone else hurts. You plot and scheme for ways to become a greater part of their lives, even if you know it's unworkable. If you've ever been in love, I'd guess that you're likely to relate to and understand at least one of these sentences.

For me, love shattered the windscreen of my soul when I was 17. Suddenly its wind was hammering into my face, filling my passages, gluing my eyelids open and making my mouth dry. I revelled in it, giving in to it totally. Over a six-month period, a primate observer (had they built a hide in my room) doing time samples of my general behaviour would have calculated the following. 1. Time on phone: up

500 per cent. 2. Time writing (poorly drafted letters): up 700 per cent. 3. Time doing sit-ups: up 500 per cent. 4. Time inspecting hair: up 400 per cent. 5. Time doing homework: down 600 per cent.

She occupied almost every waking thought. I rearranged my whole social group to get closer to her. Without even trying, I learned on what days and at what times we might cross in the school corridor. I longed for her breath, or a little hand-holding, and even attempted to snog her for the entire length of *Jurassic Park 2*. This was totally and wholly *not* me. I love *Jurassic Park 2*. I was someone else. I was infected. It infested me. Yes, I'm still in love now (with the same woman, no less), but that . . . that was really something. So powerful was this feeling that I actually seemed to stop eating at one point. Love-struck. Love-sick. Struck sick.

Perhaps you remember such feelings yourself? I'm sure you do. What often makes me smile is when people assume that this is a cultural thing. That love is a Western invention. As if culture can make you do such stupid things, can make you crazy, and *feel* things so intensely. That culture can magic up hormones and spontaneously enchant your genitalia whenever they're around. Nope, I don't buy that. It's deeper than that. This is anything but imagined or copied from others. And many of the feelings that we associate with love are universal in humankind.

All of us exhibit predictable, physically observable changes in hormone levels when we fall in love. Among the cocktail of mind-altering hormones pumping through our systems during these early stages of love is one that you may have heard a little about. Oxytocin − the so-called 'love molecule'. For almost a decade, oxytocin has been touted in the media as a wonder drug − a chemical produced by the body when we cuddle, kiss, hold hands, stroke, tweak and orgasm. It makes you *feel* good, we're told. It's a social pick-me-up; a chemical we crave - a reward or a chemical treat that encourages blind sociality and passionate bonding between two individuals. Produced by the hypothalamus, oxytocin is certainly a powerful ingredient in love's chemical cocktail. Production of it spikes when we're sexually aroused, and it also involves itself in the female reproductive system, particularly during birth and breast-feeding (other activities where close bonding has adaptive advantages). Though the exact mechanisms through which oxytocin works are yet to be fully understood (though you'd never guess that to read the headlines), we can say with certainty that it is one of a suite of neuro-hormones that our brains produce when we feel emotions of love. Looking at fMRI scans of the brains belonging to people afflicted by love's spell, you can see the brain regions associated with emotion and reward light up like Christmas lights, blood pumping, hormones working.

Seratonin is another interesting ingredient. Rather than rising, though, levels of this neuro-hormone drop when love takes us, falling to levels more commonly seen in those with obsessive-compulsive disorders, which perhaps isn't surprising; love leads to nothing if not single-mindedness, after all.

Let us turn to other animals. Could these chemicals, if they appear in other animal brains, tell us something of love elsewhere in the animal kingdom? Predictably, yes, though perhaps 'maybe' might be a safer word to use.

A *cause célèbre* of science's 'oxytocin revolution' is the prairie vole. It is a small rodent, a resident of many North American grasslands, and it looks pretty much like any other vole. But it is of immense interest to scientists because, like the jackdaws, it's a lifetime pair-bonder, though unlike the jackdaws prairie voles do engage in the occasional bit of extra-marital sex. Normally, though, they huddle with their life-partners, they groom one another, they share their nests and they raise pups together. They are among just a handful of mammals to do so (mammals, unlike birds, don't go for monogamy much).

And what of their brains? Are they like ours – hormone-filled – when in love? Interestingly, predictably and resoundingly: yes. Oxytocin and other neuro-hormones (namely vasopressin) are there, all present and correct, pumped out during sex and other acts of closeness. Like in us, in prairie voles acts of love are rewarded by the brain. Get close with a suitable mate, and your brain says the chemical equivalent of 'well done: here's a treat', and then pats your head. Like a dog that hangs around the biscuit tin, the male and female are sucked in. In prairie voles, monogamy emerges.

Not convinced that chemicals are the key yet? Listen to this. In some studies, scientists have actually managed to manipulate the brains of prairie voles artificially, to see what happens to their behaviours once love's chemical cocktail has been adjusted. They've blocked the absorption of vasopressin, for instance, to see what happens. The result? It's almost laughable: the voles become (almost) eponymous love-rats. Uninterested in monogamy, in other words.

Is this like human love, I hear you ask? Not in a literal sense, no. But . . . it's something. Something addictive. Something attractive. Something rewarding that those prairie voles know. It sure looks familiar. It has many of the hallmarks of love, I suppose you could say.

Perhaps a bigger question, though, is why animals sometimes evolve a monogamy reward system at all. After all, why would Richard Dawkins's selfish genes, obsessed with spreading themselves far and wide, favour a strategy that appears to oppose the very concept of spreading? If genes are your currency, in what situation would it pay to sit at home, with the same partner, night after night? If you have legs, why not get out there and sow your genetic seeds elsewhere? You'd pay out, genetically speaking, surely?

So why? Why monogamy? Well, it seems that the honest answer to that question is this: monogamy doesn't pay, not normally anyway. For nearly every animal life form on

Earth, monogamy is an evolutionary dead-end. It doesn't work. Some animals might dabble in it, sure, but many will ultimately fail. The genes for love-rats (male or female) will outcompete you, outmanoeuvre you and ultimately extinguish you. They will leave you literally and metaphorically holding the baby – your genetic investment squandered. Drowned out.

So why does monogamy even appear at all? Bluntly, it occurs when there is no other choice. Monogamy tends to emerge only in species that require two parents for the successful rearing of offspring. If one skips out on its responsibility and seeks mating opportunities elsewhere, the offspring die and along with them go the philandering genes. For these species, monogamy becomes the only game in town, and genes that promote monogamy (including those responsible for a monogamy reward system, *i.e.* love) become the only ones to flourish. But this is a cheap and dirty popular description of why monogamy pops up in life's tree. Increasingly, scientists are positing other factors being at play in its emergence.

In recent months there have been two papers on the evolution of monogamy, both of which offer up new theories as to what may be a driving force behind its evolution. One says, in modest terms, that monogamy pops up when females are distributed widely across a habitat, since males can't evolve to handle more than one at the same time. It ends up paying the males (in terms of fitness) to stay close by a female rather than run around all over the place seeking mates, only to have their breeding partner (and genetic legacy) cuckolded while they're away. If you don't want to be cuckolded, says this theory, stay close (in some ways I suppose this is what the non-commensal mice in Chapter 12 are up to). In the other paper, which focused exclusively on primates, monogamy is hypothesised as a way to limit infanticide, specifically the killing of offspring by an intruding male while the father (more often than not) is off trying to sow his seeds elsewhere. In this situation, everyone stays at home because they're terrified of losing the genetic silver to an incoming sexual competitor with bloodlust.

It's important to say that these two theories aren't necessarily mutually exclusive, either – it can be one or the other or both. Or neither, of course. And the factors at play behind the evolution of monogamy in primates might not be those responsible for monogamy in prairie voles or jackdaws.

Despite all the genetic risk involved, it appears a bit of a miracle that monogamy should ever have evolved at all. Yet it has, relatively sparsely, over a host of species. Examples of monogamy in the animal kingdom are dotted throughout the textbooks, and some examples are well known and even (rightly or wrongly) revered by well-meaning righteous folk; among the best-known examples is *March of the Penguins*, the Oscar-winning documentary, which received glowing adoration from various oddball commentators for showing humans the hard-won benefits of monogamy, despite the fact that emperor penguins often choose a different partner each year.

Among the *other* famous monogamists is an antelope called Kirk's dik-dik, males and females of which form tightly bonded pairs that roam Africa's eastern and southern heartlands. And why does monogamy seem to work for them? Some suggest that the male hangs around to mask out her smell from other males, but really no one seems quite sure just yet.

The monogamous fishes include the convict cichlid. Males and females of this species rear young together, in their own little crevice at the bottom of the lake, protected from predators and intruding competitors. Perhaps these crevices are nature's sex architecture: a fortress to keep out the cuckolders? We're still waiting for answers on that one, too.

Among the reptiles, arguably the best-known mono-gamists are the shingle-backed skinks (often called the

Australian sleepy lizards). These are big, floppy reptiles that are built like tanks. You often see them lumbering around the outback in pairs, almost joined at the hip. In one study that tracked the lizards for up to five years, only 18 per cent of males mated with a female that was not their mate. For monogamists, these are impressive statistics. For reptile monogamy, more so. And what do these lizards get out of it? One theory is that the gravid (pregnant) female shingle-backed skink benefits from having an extra pair of eyes to look out for predators. And him? Perhaps his benefit comes from the fact that, by being around, he's less likely to see his spouse get eaten before she gives birth to their shared genetic offspring. Everyone's a winner, then.

In birds, perhaps the most celebrated monogamists are the albatrosses, in which pair bonds stand the test of long months at sea, and years and decades can pass in apparent contentment. The reason for their monogamy is pretty simple. On a cliff top, where aerial predators and competitors abound and your offspring are at risk of falling to their death at any given moment, someone has to look out for baby. The adults take it in turns. Monogamy pays. Even here, though, extra-pair copulations occur, albeit more rarely than in other 'social monogamists'.

But the term 'monogamy' is a loose one, because any animal that mates only with one partner over a lifetime or a season can be described as such. With this loose definition some insects can also be considered monogamists, including the housefly, the one that drones around the kitchen lights in summer and later that night appears in your bedroom, raising your blood pressure frustratingly just before lights-out. They mate only once, and with one partner.

But jackdaws are different to every single one of the animals I've listed above. They are sexually monogamous, not socially monogamous. Truly monogamous, in other words. Year in and year out they're there, hanging around with the same old partner. Properly monogamous. They are freaks, really. Outliers on the monogamy spectrum. Even if it is one day proven that they *do* occasionally invest in extra-pair liaisons (and to be honest I'm sure some probably must). I would be happy to see us continue to bandy about their title: super-heavyweights of the super sexual monogamists.

And so we come to perhaps the biggest of all questions. What is it about jackdaw life that has led it down this ne'er-trodden evolutionary route? What made them super-monogamists? Well, at the moment it seems your guess is as good as mine. Perhaps it has something to do with the social (and spatial) organisation of jackdaws? Or with their mighty, intellectual, questioning minds? No one appears yet sure. Maybe the answer will surprise us.

As I write these words it has been three weeks since I ran over that jackdaw, knocking it to the ground in front of its partner. It still fills me with an aching sadness that I'm embarrassed to admit, and which I occasionally relive. I have taken to talking about it with bird experts, and have been met with similar such stories, each bookended by bouts of ringing grief. Monogamy roadkill: there can be few things more heart-breaking on planet Earth.

As some sort of therapy, and on the advice of some of these experts, I've chosen to drive to Cambridge to see one of the greatest expressions of love in the animal kingdom, or so I've been told. I've decided to make a pilgrimage to the Madingley Jackdaw Roost. I have no idea where exactly this roost is in Madingley, but according to almost everyone I've spoken to about it, it's a late-season spectacle not to be missed. You turn up, see thousands of jackdaws, and then go home happy, or so I'm told.

Madingley is a pleasant little village, full of chocolate-box houses and littered, charmingly, with 'TOAD CROSSING' signs. In between searching the sky for jackdaws and keeping an eye out for toads I manage to somehow keep myself from putting the car into a ditch or into the path of oncoming traffic. Ten minutes pass as I drive around. Then 20. After driving around for 30 minutes or so, I decide my search for jackdaws is hopeless. Instead, I park up and wander the streets on foot, hopeful that at some point I may hear them passing overhead or see them sitting in the overhanging trees. Bracing myself against the biting wind, I trawl through never-ending piles of crunchy brown leaves and long-forgotten conkers. The daylight drifts as I make my way through the village from east to west, and there, in the deep distance over the noise of the A14, I can hear it – the jackdaw calls ringing out, a barking cacophony. Forwards.

In the centre of the village sits Madingley Hall, with its rich lawns and large pond. It cuts a hole in the woodland that surrounds the village, and provides a large frame through which I can finally see the sky. And there, up there, finally, I see my first jackdaws of the day. Small parties of them move high overhead, heading west. I stand with my eyes skyward, my head at 90 degrees. Three, four, nine. More and more appear, all silently moving westward. Fifteen, twenty-five, forty. A flock appears. Then another. Then more. After minutes of this my neck begins to ache. I carry on peering skyward, mouth slightly open. Three small flocks appear to coalesce and form a larger, self-organising gaggle of at least 50 or 60 in the red sky. And then I spot a trend. The most amazing thing is this: even from down here I can see that many are flying in pairs. Even in a flock, I can just about make out the invisible string that tugs at some of the couples, forcing them to gravitate back toward one another like binary stars in a swirling galaxy, a galaxy where the sky is white and the avian stars are black as night.

Jackdaws flock throughout the year, but as the wintry easterlies set in they take this behaviour to ever greater heights, dizzyingly so, sometimes forming enormous super-roosts like those that occur here at Madingley. Although not the largest in Britain, the Madingley roost goes back centuries, almost to the Domesday Book. Up to 10,000 jackdaws turn up here on winter nights (some travelling 150 miles for the privilege). They spend an hour or so spiralling, spinning, surfing the twilight before heading to a given roost site, usually a collection of old trees to the west. Before collecting for their main roost at dusk, pre-flocks form in the village and the surrounding farmland, made up of hundreds, sometimes thousands, of jackdaws. Yet, here I am. There's none on the ground or in the trees, and I can only hear their faint cawing as they continue to trundle overhead, those gravitating pairs with their measured flaps.

I've been told that they carefully monitor one another during this, the off-season for sex. Their pairs remain strong. They keep tabs on one another, looking out for their spouse as need requires. They continue to be drawn to one another even though sex is a long, long way off. The planet still has to circle around to the other side of the Solar System. They don't seem to care.

Is this what love looks like in jackdaws? Are they like the prairie voles? Do their brains light up like ours? Are they hormone junkies like us? After researching the topic for months, last week I stumbled upon the right research paper, published, like the monogamy papers, only recently. And the results are as predictable as they are interesting. Where once the neuro-hormones of attachment were considered a uniquely mammalian trait, scientists now think they may have found something similar in birds, or rather in the brains of the (monogamous) zebra finches, rather than jackdaws (we await someone to look under their bonnet, so to speak). According to the research, a familiar story is at play in these birds. The paper reports that neuro-hormonal activity associated with pair-bonding has been found, along with an 'oxytocin-like' reward molecule, like those that we mammals know so well. Indeed, if you inhibit the action of this 'oxytocin-like' molecule in these birds, then, hey presto, the zebra finch's monogamous behaviour is affected, specifically pair-bonding and allopreening (the bird equivalent of combing your partner's hair). Like prairie voles, there is a reward mechanism of some sort involved here. And who knows, perhaps the same mechanism is at work in the jackdaws? Perhaps that's what keeps them together?

It's early days, but few would disagree that this research could be a fascinating glimpse, perhaps, of a form of convergent evolution in mammals and birds; of neural systems that reward monogamous behaviours with feelings that we (or shamelessly I, at least) might refer to as love. Who knows, maybe their dinosaur ancestors also knew of it? The thought of tyrannosaur heartache is enough to make your inner child blush. Once more, perhaps science is about to smash yet another pedestal. Consciousness, theory of mind, love . . .

I walk to the edge of the village and on the far side of the fields beyond I see what I've been waiting for. At last, they're there. Hundreds of jackdaws are scattered across the trees in front of me. They hop. They dally. They sit. They watch. This is a larger pre-roost – much larger than anything I have so far seen. A flock readying itself for a final movement to the big roost in the woodland west of the village. Some jump sporadically into the air. They lean and lurch among the branches. Some circle, some ride the currents above the trees. Most sit still, bored. Another hundred or so jackdaws sit in the middle of the field, not really up to much (many jackdaws seem to spend most of their days doing very little at all). Some at the edge are larking around, cawing and yapping among themselves. In the fading light their dark shapes become like black bulbs yet to be lit. Here on the ground some appear to be paired. I climb up the nearby gatepost and perch myself tentatively on the top. I take them in for a couple of minutes, my frosty breath thick in front of my face, carried away by the increasingly sharp wind.

Suddenly, and without any obvious sign of warning, all of them – every single one – take flight. Within seconds they are all completely airborne. Perhaps 500 jackdaws ride, loop and swing as if on threads. They lift and hover against the breeze and, as a collective, seem to hold themselves there. Then, like paratroopers leaving a plane, the wind seems to take each one. One by one they shrink into the distance. Bit by bit they go, their monotonous yakking fading into the almost-night. I look around the sky now, and suddenly they're everywhere I look. Small flocks, big flocks, couples, quartets – pairs of pairs of pairs of pairs. There's too many to count. Far too many. Hundreds, thousands it seems – their jabbering like a raging ocean. Applause from all sides. All of them heading west, all in their own time. All of them heading outward and over to their winter roost, to sleep together.

For 10 minutes or so I bask in the spectacle. I let the light fade around me, still sitting on that fencepost, even though I can no longer feel my buttocks. With my collar up, hood on, hands in pockets, I smile broadly. If I listen carefully I can still hear them, like distant frenzied applause on the horizon. A few still pass high overhead. Dribs and drabs. They drift downwards and toward the horizon, like lost honeybees; some as small as iron filings, pulled magnet-like toward the setting sun. Before long, I'm sitting alone with my eyes closed in the dark. I continue to smile.

Sex is months away for these jackdaws, and yet . . . just now, there they were, most of them paired up, faithful, ready for winter, trying to stay alive long enough to have sex again, should the seasons play ball. Like so many stories in my journey, sex sits there in the background, organising the daily lives of Earthlings – all of us – like an invisible force. Like gravity. Like a magnet, then.

I suspect that if we re-ran life once again from a single cell billions of years ago, the faithful jackdaw may not turn up again today – like language, theory of mind, tool use, but perhaps even rarer. The proper monogamists, fleeting manifestations of the truest of loves; an adaptation that came courtesy of a universe that cares nothing for their fate or their survival, but that has stumbled, if only momentarily, upon a picture of sheer faithfulness, a mode of life that we humans can only watch and try, if we fancy it, to replicate. Sitting there on that fencepost, the winter setting in, I could think of no better end to a journey that we all travel and that we all come to know so well. 'Till death do us part'; the most unimaginable but imaginative expression of selfish genes that one could ever have dreamed. Love: the most fantastic feeling that any of us can experience, an unexpected and surprise accolade from a planet seemingly obsessed with self-interest and one-upmanship; a magical evolutionary fruit that only a few animals may ever come to know. That you and I know. 'The autumn wind sings the song of the elements,' says Lorenz. And how sweet it is.

Epilogue

I started this book with those pandas. Or rather, I started this book with a panda's bum. That was a year ago, almost to the day. I was a panda apologist back then, uncomfortable with the popular notion of them being cack-handed when it came to sex. It spurred me on to this, my foray into the sex lives of animals. And now, we're at the end.

Perhaps now might be a suitable time for me to update you on what happened to the owner of that monochromatic rump at the start, Edinburgh Zoo's Tian Tian. Firstly, I can't lead you on. It didn't work out. They didn't breed. No panda babies, at least this year. But it was looking very promising for a while, and it really did turn into what felt like a daily soap opera for a month or two over the summer. In the build-up, through a series of press releases, Edinburgh Zoo reported that all the right moves were being demonstrated. Tian Tian was reported to have become 'grumpy' (the hormones, don't-cha-know). Yang Guang was adopting the handstand position to get his scent out there, ready for the big show. But then . . . nothing, alas. In Chapter 6 (when I went to see those horses have sex for money), it was announced that Tian Tian was unlikely to conceive naturally this year, for reasons that we were never really privy to. The experts couldn't let her fertility window go to waste, though. She was anaesthetised and artificially inseminated with defrosted sperm from her zoo-mate Yang Guang and from Bao Bao, the dead panda from Berlin Zoo. The 'panda debate' continued in the media, and sex continued to be at its heart.

'Stuff the panda!' argued experts on BBC2's *Science Club*. 'They're not helping themselves!' they shouted. On Radio 4's *The Infinite Monkey Cage*, another expert declared, 'Pandas deserve to be extinct – they're reproductive once every 20 years or something, and only if they're in the mood!' 'Pandas struggle to procreate because of low sex-drives and infertility issues,' said one idle sentence in the *Daily Mail* (and 1,200 other web pages that copied and pasted the text into their own documents as fact). 'The animals are notoriously difficult to breed, often lacking a sex-drive and sometimes accidentally crushing small cubs shortly after birth, leading to falling numbers in the wild,' says another *Mail* article. You can't help but sigh.

Some felt sorry for Tian Tian. 'How awful to be fertilised in such an invasive way, and then for it to fail,' they said. 'We should let them face extinction with dignity,' said others. I was torn. I still am. For panda-lovers and others who court this panda circus sideshow, isn't it right to use the technology at our fingertips, the scientific knowledge gleaned from years of research to 'save the species from extinction'? We have the know-how to make creatures reproduce – why shouldn't we use it? An ethical, almost moral question, indeed, and one I'm not sure there's a good answer to, or one that I could provide you with (then or now, as I write these words at the end of my journey).

During the press coverage surrounding Tian Tian's artificial insemination, social media was alight with all sorts of boorish controversy. 'It's time we let nature say goodbye to these creatures and focused scant ecological resources on species that at least meet you halfway!' they said (halfway? Meeting us *at all* was the worst thing that ever happened to them).

So what happened to Tian Tian after her insemination, you might ask? Well, let me tell you. For a while it sounded positive. On the 9th of August the zoo's press release said the following:

Although it is still early days, the Royal Zoological Society of Scotland can reveal that we are not ruling out that female panda Tian Tian may be pregnant . . . A second hormone rise in progesterone levels was detected in Tian Tian on 15th July, and then confirmed on Wednesday 7th August, which indicates she may be pregnant or experiencing a pseudo pregnancy; this means that in around 40 to 55 days Tian Tian will either give birth to a cub or her false pregnancy will end. If there is a cub, it could be born between late August and early September.

There then followed speculation each day, spread over a number of weeks, about whether Tian Tian was or wasn't pregnant. Rarely does nature write such media-friendly scripts. After weeks of wondering, on the 15th October 2013 the world got its answer. Chris West, the Royal Zoological Society of Scotland's CEO, said:

Such a loss has always been in our minds as a very real possibility, as it occurs in giant pandas as well as many other animals, including humans. Our dedicated team of keepers, veterinary staff and many others worked tirelessly to ensure Tian Tian received the best care possible, which included remote observation and closing the panda enclosure to visitors to give her quiet and privacy. We are conducting a detailed review of the scientific data collected, but I am totally confident that we did everything it was possible to do.

A sad end, for this year at least.

I think back to my trip up to Edinburgh Zoo late last year. I half expected back then that I would revisit the zoo around about now. I imagined writing a jovial piece about Tian Tian with her cub, about how sex science was booming, and immense and exciting changes were afoot in the world of sex. Yet here I am now, at home, as the snow starts to fall.

A number of things struck me while writing this book. One in particular. As I drafted these chapters I found myself marvelling at how much we don't yet know about sex. How did sex begin? Does an Australian sleepy lizard feel love like a bird? Does a bird feel love like a vole? How can we get rid of pesky Spanish slugs? Where are the male flowerpot snakes? How much does ocean noise affect the sex lives of aquatic mammals? How do slug mites have sex? How big was a *Tyrannosaurus* penis? Who made that plug in the hedgehog's vagina? Why are dolphins often as homosexual in their tastes as heterosexual? Do glow-worms have sex with streetlights? You get the idea. We've had the Darwinian revolution. The Lorenzian revolution. Perhaps this century we will see a similar sexual revolution. A *proper* sexual revolution. It strikes me what an incredible time it is to be alive, with questions like these still to be answered. We have much to look forward to.

Yet in other ways we still appear so backward-looking. George Levick, the Antarctic explorer who was so fearful of the academic response to his observations about Adélie penguin sex, lived a century ago. Occasionally I have wondered how far we've come since then, in our public understanding and discussion of such issues. The science of sex-acts such as masturbation and homosexuality are still in their infancy, partly because scientists fear that their results won't be taken seriously. Penis gags litter the popular press, yet vagina stories rumble past like pubic tumbleweed. Nemo is anything but a sequential hermaphrodite, according to Disney and Pixar. Pandas are anything but competent masters of sex in the wild, according to popular opinion. If there is a scientific revolution taking place, it will have to work hard to attach itself to a cultural revolution that continues to drag us back to Levick's time.

I wonder whether the frown-inducing headlines that we see about pandas are something about which future generations may chuckle and roll their eyes. I remember fondly that medieval description of a hedgehog that Hugh Warwick showed me: 'That the Hedge-Hog is a mischievous Animal; and particularly, that he sucks Cows, when they are asleep in the Night, and causes their Teats to be sore.' Perhaps we might end up looking back on our panda headlines a little like this? I hope so. As I said at the start, and just to reiterate, you really can't fault panda sex lives, not wild pandas anyway. After all, their ancestors have the same batting average as you: they've never missed a ball. Same with almost every animal you care to name. One hundred per cent. It's a tie, in other words. After 12 months of reading and writing about animals having sex, I'm as convinced of this simple fact as anything else. Everything is a master of sex. Equally.

And that includes us. Yes, you and me. The elephant in the room. You'll notice that I have largely avoided addressing the sex lives of our own species, and for two good reasons. Firstly: I don't consider us a big deal. This is, after all, a book about sex on Earth, and we're but one (particularly invasive and adaptable) resident of millions. But there is another reason I've kept largely quiet about us, and that's because I don't think we know enough about ourselves (in an evolutionary sense) to brandish ourselves as 'THIS' or 'THAT'. There are too many missing jigsaw pieces yet to be found.

In the one chapter in which I came close to bringing humans up in any depth (the one about jackdaws and love), an early reviewer of the text (to whom I am enormously grateful) expressed understandable concern: 'Your argument about monogamy sounds a little like having two parents is better than one,' he wrote in the margin. 'Could this be used to criticise single parents?' he scribbled underneath. He was right to have concerns. After all, the well-trodden argument goes that humans are monogamous, and monogamy generally appears in animals (especially birds) where child-rearing costs are high – it takes two, in other words. Therefore, the argument appears – monogamy exists in humans because it helps to rear offspring – single parents be damned! 'Failures!' say the hard-liners. Let's all do the decent thing and stay with our partners, the more conservative among us say. Divorce is killing society, others might argue. But hold on, hold on. Listen up, because this bit is important: did I say in this book that humans are monogamous? No. I said that we have a neuro-hormone system that gives us feelings of reward when we are with the object of our affection, something found in social monogamists like prairie voles and zebra finches. We evolved a similar system, which might suggest we were drawn at some point toward monogamy, but then we *also* evolved a penis shaped like a plunger to suck out the sperm of rival males. Go figure.

Perhaps one day we will discover what humans truly are (or were), but now, in the modern world, we are anything we want to be, or at least we should strive to create a world where this should be so for all people. Sometimes, we just need to get over ourselves.

In that chapter about monogamy, I gave a passing mention to *March of the Penguins*, the 2005 feature film by Luc Jacquet.

It depicted the lives and loves of Antarctica's emperor penguins. You're no doubt familiar with the story of their sex lives, even if you didn't see the film. Male and female emperor penguins make an epic journey each year, waddling from the ocean inland to their ancestral breeding grounds. They undertake their courtship routines, they mate, and then the female lays a single egg. For the chick to survive the male and female must take it in turns to stand there, on the freezing ice, huddled up against the coldest winds imaginable, protecting their young while their mate goes off to find food. The film was revered by nature-lovers, and won multiple awards (including an Oscar for best documentary). And it was lauded by others, too, some with questionable intentions. Some saw in it social commentary; a deeper message. Advice from penguins to humankind. They saw in it a film that promoted 'conservative family values' or 'a metaphor for family values – the devotion to a mate, devotion to offspring, monogamy'. Even the word 'self-denial' was mentioned in one article. It became an international talking point. 'Just what was Jacquet trying to *say* with that piece?', people asked. They went over it frame by frame, pursuing a deeper message and meaning in everything. When asked about it, Jacquet attempted to quash this suggestion on numerous occasions, arguing he was simply making a film about nature – nothing more, nothing less. And, as the questions continued, at one point he used a line that I rather liked for its concise eloquence. I like to imagine him drawing breath before saying it, taking a moment to gather his composure before calmly laying it straight to the journalist questioning him: 'You have to let penguins be penguins, and humans be humans,' he said simply.

And, do you know what? It really is as simple as that. A sentence that applies today in the same way that it did to George Levick, sitting cold and alone at the bottom of the world just over one hundred years ago, watching in horror as those Adélie penguins copulated with dead females, with chicks and with the rocks beneath their feet.

Sometimes, Levick old chap, you just have to let penguins be penguins. You have to let jackdaws be jackdaws. Let hedgehogs be hedgehogs and snakes be snakes. Let frogs be frogs, let dolphins be dolphins and let pandas be pandas. The point of all of these creatures is nothing to do with us. Their point is only to make more. To copulate. To breed. To replicate. And how much more wonderful life on this planet is as a result. Sex made us special. There may be no other place where it occurs in the universe. Each and every act is something wonderful, in its truest sense. A reproductive quirk from which few of us ever escape, but of which everything currently alive on this planet is a master. You, me, the penguins, the frogs, the snakes, the hedgehogs, the slugs, the slug mites. And there, as near to the top as everything else, are those pandas: resplendent, knowing, perky and committed . . . and smelling fantastic (if you're a panda, that is).

Sex is the key to life's past. And it's the key to our future, with or without the pandas. With or without us. Best enjoy it while it lasts.

Acknowledgements

It must be bloody annoying to live with someone who has to keep rushing out to speak to sex scientists, or to track down horny toads and other animals that show no respect for things such as diaries or dinner parties. So it seems right for me to begin by acknowledging Emma, my wife, to whom I am eternally grateful for her lack of rage, anger or resentment. Thank you for your encouragement and helpful words throughout, from start to finish. You were never anything other than positive. I love you. Thank you.

There are a host of other people who I must also thank, and without whom this book would have just been me, commenting on YouTube videos of animals doing it. In no particular order these are: Henry Nicholls (author of the excellent *Way of the Panda),* Sharon and Peter Flint (of the Flint Entomological Consultancy), Patricia Brennan (University of Massachusetts Amherst), Martin Wikelski (Max Planck Institute for Ornithology), Terry Whitaker (lepidopterist extraordinaire), Chris Wilson (Imperial College London), Helen Smith (of the British Arachnological Society), Tim Strudwick (RSPB), Mark Simpson and Rebecca Lee (WWT), Martin Fowlie and Ade Long (BirdLife International), Paul Hetherington and Alan Stubbs (Buglife), Sally Bate (horse vet extraordinaire), Anita Joysey and David Seilly (The Wildlife Trust BCN), Iain Barber (University of Leicester), Ian Bedford (John Innes Centre), Hugh Warwick (author of *A Prickly Affair* and all-round spokesperson for hedgehogs), Ruth Kent (my faithful cuttings editor), Hannah Urpeth (mites), James Dunbar (fen raft spider video-maker), Abi McLoughlin (Bat Conservation Trust), Becky Wragg Sykes (palaeolithic archaeologist) and Matt Hann (communications magnate). Many of these people read through early chapters (and some read the whole thing), putting me straight where I needed it. Thank you.

Enormous thanks to Sam Goodlet (www.samdrawsthings. com) for the wonderful chapter illustrations. Thanks also to the Edinburgh Zoo staff with whom I spoke on my visit.

It seems fitting that I also single out an important person who I have never actually met, but without whose book I would completely have failed to gauge properly the multitude and magic of animal sex lives. She is Olivia Judson, and her book is *Dr Tatiana's Sex Advice to All Creation*. It is brilliant and well worth a read.

My thanks, of course, to Bloomsbury, particularly Jasmine Parker, Vicky Atkins and especially Jim Martin (who called me 'MR SEX' throughout, which I found very off-putting in meetings). A big thank you to Jane Turnbull (of the Jane Turnbull Literary Agency) and Jennifer Waterman at Arlington Talent Management, too.

Spending lots of time, alone, writing about animals having sex has the potential to drag a writer into some sort of mild fugue state that rings, ever so slightly, with a hint of questionable perversion. My friends and followers on Twitter have been a great source of insight and general loveliness – and only a few chose to un-follow me when they saw the infamous picture of the pixelated echidna penis. Thanks to you all. (And please do keep sending me your pictures of animal penises. The joke hasn't got at all tired yet.)

Lastly, I wanted to say something else. In an early draft of this book, there was a chapter about my mother, who had a long career as a psycho-sexual therapist and educator. Having discussions about sex with your parents is always rather heady, but my mum probably somewhere, somehow, made it interesting enough that I should never hide from it, or let sex dog me, so to speak, throughout teenager-dom and into adulthood. Their support (for my dad was also comfortable with the subject), particularly in the last decade, has been nothing short of heroic, and we are all so grateful. Thank you both. And, well . . . should I also offer them, my mother and father, a thank you for their own sex acts? Particularly the one that created me? Eww. No. That would be weird. But now, dear reader, you understand why this early chapter about them had to be completely removed. And burned.

Jules Howard, April 2014

Index